ALGEBRA
A Book for Adults

ALGEBRA
A Book for Adults

SUZANNE K. DAMARIN

JOAN R. LEITZEL

The Ohio State University

John Wiley & Sons

New York   Chichester   Brisbane   Toronto   Singapore

Library of Congress Cataloging in Publication Data:

Damarin, Suzanne K.
   Algebra: a book for adults.

   Includes index.
   1. Algebra.   I. Leitzel, Joan R.   II. Title.
QA154.2.D34   1984   512.9   83-7023
ISBN 0-471-86274-6

Printed in the United States of America

10 9 8 7 6 5 4 3 2 1

PREFACE

In recent years the number of adult students enrolled in four-year colleges and universities and in two-year colleges has increased dramatically. The entry-level course which this book was written to support developed from a recognition that adult students are often not well served in traditional freshman mathematics courses. This course is an alternative to the remedial sequence of courses required for underprepared eighteen-year-old freshmen. This textbook has been effective in courses designed specifically for adults and in courses where adults and traditional students are taught together. It prepares students well for the precalculus courses preceding engineering calculus and business calculus.

The target audience for this book consists of students who are several years away from formal instruction in mathematics. These adult students typically took algebra in high school but have lost algebraic skills through lack of use. They are often competent in the management of everyday affairs, read regularly, and reason well to arrive at sound economic decisions. They are self-directed and have the ability to work hard. Although these students may be apprehensive about college-level mathematics, they are generally capable of college-level work and are confident of their abilities in any area where they may seek a major. Our experience is that adult students can learn faster and ultimately with deeper understanding than is frequently assumed. Furthermore, they overcome their anxiety toward mathematics as they become successful in learning it and as they see applications within their own experiences.

This book treats adult students as mature individuals and specifically provides for the building of confidence in these students. It identifies arithmetic skills used in the adult workaday world (page 4) and then builds on these skills in the teaching of concepts, techniques, and reasoning skills required by and for the use of algebra (pages 25, 48, 66, respectively). The book consciously corrects common adult misunderstandings about mathematics (page 12).

The order of topics is chosen to move students quickly to key ideas of algebra (page 6) and to delay numerical complications. Initially, variables are defined over the domain of natural numbers; principles for simplifying algebraic expressions and for solving linear equations and inequalities are then developed over this domain (page 33). Students are quickly immersed in the applications of algebra where they are confident and competent (page 2).

Later the number domain is expanded to include integers, rationals, and reals (pages 48, 93, 171, respectively). Algebraic techniques are extended to these systems (page 60).

The focus on applications drawn from the adult world is retained throughout the text (page 186).

The practice of arithmetic skills is an intrinsic part of the problems and applications throughout the text (pages 10, 147). In addition, four appendixes are provided for student self-instruction in computational topics. A Pretest in each appendix guides the student's self-directed study (page 221).

The book encourages the use of scientific calculators but does not require them (page 53). However, students are never required to do computations without a calculator.

Answers to odd-numbered exercises, and all answers to the Review Problems are provided in the back of the book for student use. Careful analyses of problems that can be approached in more than one way are given in the Teacher's Resourses. The teacher's materials also include several parallel forms of mastery tests and problem-solving quizzes for each chapter. A recommended testing procedure for adult students is described that permits tests to build student confidence while measuring achievement.

Teachers using this book for a course of fewer than 15 weeks can choose to eliminate any of these sections without interfering with other topics: 3.3, 3.4, 4.6, 4.8, 6.4, Chapter 7, Chapter 11, and Chapter 12. Chapter 12 is independent of Chapter 11; teachers may select topics from the last two chapters to complete the course in a way most appropriate for the course structure at their own schools.

SUZANNE K. DAMARIN
JOAN R. LEITZEL

ACKNOWLEDGMENTS

Many people contribute to the development of a book, yet not all of them can be thanked individually. As we recall the evolution of this book, we would particularly like to acknowledge the support of Vice-Provost Arthur Adams, Dean Colin Bull, Associate Dean John Riedl, and former department chairperson Professor Joseph Landin, all of whom supported the initial development of a mathematics course for adults at The Ohio State University. Conversations with many students and teachers of the course have contributed to the current book; we note, in particular, the contributions of Judie Monson and Janeal Mika, both superior teachers of mathematics. Dodie Shapiro not only typed the several drafts of the manuscript, but also applied her own adult understanding of the mathematics to the drawings and graphs in the book. Our husbands and children provided consistent moral support and also occasional help with indexing, proofing, and other details. Robert Pirtle, our editor at John Wiley & Sons, quickly perceived the unique features of this book, and its importance at this time, and guided it to completion. Finally, we wish to acknowledge that we are among the many teachers of mathematics who have been greatly influenced by the values and skills of a master teacher, Arnold E. Ross.

S.K.D.
J.R.L.

CONTENTS

ALGEBRA
A Book for Adults

CHAPTER 1

THE LANGUAGE OF ALGEBRA

1.1 STARTING WITH ARITHMETIC

We begin our study of algebra by carefully examining some specially chosen problems in arithmetic. Since algebra is an extension of the familiar ideas of arithmetic, this is an appropriate starting point.

The telephone directory gives the charges for a long-distance station-to-station call from Columbus to Chicago as $1.95 for the first three minutes and $0.32 for each additional minute or fraction of a minute.

This information enables customers to compute the cost of a call from Columbus to Chicago if they know how long a conversation will last. Some of the possible costs are summarized in the following table.

Number of Minutes	Cost of Call
1	$1.95
2	1.95
$3\frac{1}{2}$	2.27
5	2.59
$5\frac{1}{2}$	2.91
6	2.91
$6\frac{1}{4}$	3.23

One thing the table demonstrates is that a person who wants to limit the cost of the call to $3 can talk up to 6 minutes; a call of more than 6 minutes will cost more than $3. (How long can you talk on the phone if $5 is all you can afford?)

Commonly, one quantity can be computed in terms of other known quantities. Here are some additional problems for examination and discussion.

PROBLEMS FOR EXAMINATION AND DISCUSSION

1. Two travelers start from the same place at the same time and go in opposite directions, one at 15 miles per hour and the other at 25 miles per hour. If the time the two have traveled is known, then the distance between them can be computed.

Time Traveled	Distance 1st Traveler	Distance 2nd Traveler	Distance Apart
1 hour	15 miles	25 miles	15 + 25 = 40 miles
2 hours	30 miles	?	30 + ? =
3 hours	?	?	?
4 hours	?	?	?

When the travelers are 240 miles apart, can you say how long they have been traveling?

2. Fruit drinks similar to Hi-C and Hawaiian Punch are made by combining pure fruit juice with artificially flavored water. Whether a mixture of juice and water can be legally labeled "fruit juice" or must be called "fruit drink" depends on the percent of juice in the mixture. Suppose a vat contains 10 gallons of a 40 percent juice mixture. This means that it contains 4 gallons of pure juice and 6 gallons of flavored water. If more water is added to the vat, the mixture will be diluted and the percent of juice decreased.

Number Gallons of Pure Juice	Number Gallons of Water	Total Number of Gallons	Percent of Juice in Mixture
4	6	10	40%
4	8	12	$33\frac{1}{3}\%$
4	12	?	?
4	16	?	?

How much water must be added to the original juice mixture to get a 10 percent mixture?

3. There are many rectangles that have a perimeter of 48 inches. If the length of a side of one of these rectangles is known, its area can be computed.

Length	Width	Perimeter in Inches	Area in Square Inches
18	6	2(18 + 6) = 48	18 × 6 = 108
20	?	2(20 + ?) = 48	?
16	?	2(16 + ?) = 48	?
14	?	2(14 + ?) = 48	?

Which rectangle with a perimeter of 48 inches do you think has the largest area? (Perimeter is the distance around a closed figure. Perimeter and area are discussed in Appendix A.)

4. A bank charges its checking account customers a 50¢ monthly service charge and also a 10¢ handling charge for each check returned with the monthly statement. If the number of checks returned in a given month is known, the charges for the month can be computed.

Number of Checks	Monthly Charges
4	4(0.10) + 0.50 = 0.90
8	?
10	?

If the charges one month are $2.40, how many checks should be returned that month?

1.2   WRITING MATHEMATICAL STATEMENTS

Consider again the last problem concerning bank charges. Can you describe in words how the bank determines the monthly charges for an account?

If you do this, your rule is probably something like this.

Count the checks.
Then multiply the number of checks by 10¢.
Then add 50¢ to this amount.
Voilá, the charges!

This is a general rule in that it works for any number of checks.

Using the same reasoning that led to this rule, we can take an algebraic approach to the problem and write an algebraic formula as follows.

Let B stand for the bank charges.
Let N stand for the number of checks.
Multiply N by 10¢ to get $0.10 \times N$.
Add 50¢, to get $0.10 \times N + 0.50$.
Thus, $B = 0.10 \times N + 0.50$ is an algebraic formula for the charges.

In this formula N and B are **variables.** We can assign to N the number 0 (zero) or any natural number 1, 2, 3, . . . , and compute a corresponding value for B. As a consequence, B also takes on many different values. For example, B can take on any of the following values: 0.60, 0.70, 0.90, 1.00, 1.30, 9.50, 17.20, 1037.60 (and of course many others). Notice there are many values B cannot assume; for example, B cannot take on any value less than $0.50.

Examine the following problems. For each problem find a procedure or rule that you can write in a few sentences. Then try to write an algebraic formula which reflects your rule. To do this you will have to identify variables and the relationships of the variables to each other and to other numbers in the problems.

MORE PROBLEMS FOR EXAMINATION AND DISCUSSION

1. Ohio Bell charges commercial customers a "measured rate" of 9¢ for each local call completed. How does Ohio Bell compute the measured rate bill for each customer?

2. A wholesale catalog lists two lines of men's handkerchiefs, one selling for $6 per dozen and the other for $5 per dozen (wholesale). A buyer for a men's store is about to order several dozen of each. How should he compute the cost to the store?

3. A charitable organization is selling azaleas to raise money. The members get $10 commission for the first 25 plants they sell and 75¢ for each additional plant sold. How can they compute the amount of commission due them at the end of the sales campaign?

4. A biochemist had 1000 bacteria of a certain species in a specimen bottle at noon. Each hour the number of bacteria in the bottle doubles. How can she compute the number of bacteria in the bottle at a given hour?

Exercises 1-A

1. A motorist travels 300 miles. If the rate at which he travels is known, the amount of time the trip takes can be computed.

Distance	Rate	Time
300 mi	30 mph	10 hr
300	50	?
300	?	5
300	?	4
300	r	?

2. There are many rectangles with an area of 144 square inches. If we know the length of one side of such a rectangle, we can compute its perimeter.

Length	Width	Area	Perimeter
18	$144 \div 18 = 8$	144	$2(18 + 8) = 52$
16	$144 \div 16 = ?$	144	$2(16 + ?) = ?$
12	?	144	?
8	?	144	?
ℓ	?	144	?

3. Return again to the problems in the first section.

   (a) The telephone directory gives the charges for a long-distance stat'on-to-station call from Columbus to Chicago as $1.95 for the first three minutes and $0.32 for each additional minute or fraction of a minute. Give the cost C of a call that is t minutes long.

   (b) Two travelers start from the same place at the same time and go in opposite directions, one at 15 miles per hour and the other at 25 miles per hour. Give the distance D between the two after n hours of travel.

   (c) Describe the area A of a rectangle with a perimeter of 48 inches if the length of one side of the rectangle is L.

4. The story is told that in ancient India the Grand Vizar, tired of all the conventional games in the land, called for a new game to be developed. He was so delighted with the resulting game, now known as chess, that he offered its inventor to name his own reward. The clever young man asked simply that 1 grain of wheat be put on the first square of the board, 2 grains on the 2nd, 4 on the 3rd, 8 on the 4th, and so on, until each square was covered. Describe how you would compute the number of grains of wheat that were given to the young man. (Can you believe that if these grains of wheat were laid end to end they would stretch from earth to the star Alpha Centauri and back 4 times!)

1.3 WHAT IS A VARIABLE?

Each of the problems we have looked at so far involves variables because each requires us to reason with unspecified quantities, quantities that can vary over some set of numbers. By working with these problems, and with the algebraic problems in this and subsequent chapters, you will get a better intuitive feel for variables.

It is difficult to give a precise definition for the word **variable.** Often books that use variables do not define the term. To examine the idea of variable more carefully we consider a number of definitions and then additional examples. We begin with the definitions;

the first is an informal description from *The Language of Mathematics* by M. E. Munroe. (If you are a linguist at heart you may find the first few chapters of this book both interesting and helpful in your understanding of mathematics.)

> *To graduate from arithmetic to algebra, one must introduce a new type of symbol known as a* variable. *Men have been using variables for centuries and for at least one century have been trying to define the term precisely. However, there is still some disagreement among current writers as to how one should complete the sentence "A variable is . . .".*
>
> *Very roughly speaking, the idea is as follows. To write a statement about numbers and leave one or more of the numbers unspecified, one uses some symbol that is not a standard numeral to denote each of the unspecified numbers. These "generic numerals" are called variables, and the symbols most commonly used are letters of the alphabet. (p. 13)*

Other, more succinct definitions can be found in dictionaries and some algebra textbooks. Here are two.

> *VAR'I-A-BLE, n. A quantity which can take on any of the numbers of some set such as the set of real numbers, the rational numbers, all numbers between two given numbers, or all numbers. The set of values which a variable may assume is sometimes stated explicitly and sometimes only implied. (James and James,* Mathematics Dictionary, *p. 412)*
>
> *var'i-a-ble, n. 1. That which is variable; a thing which may vary or is liable to vary. 2. Math a. A quantity that may assume a succession of values, which need not be distinct; —opposed to* constant. *cf. PARAMETER. b. A symbol standing for any one of a class of things. (*Webster's New College Dictionary, *1958, p. 941)*

Together these definitions suggest the following information about a variable.

> *A variable is a symbol or is represented by a symbol, usually a letter of the alphabet.*
>
> *A variable stands for a number from some set of numbers.*

There is some other terminology related to "variable" that needs to be clarified. In the sentence above we said, "A variable stands for a number." We may also express this idea by saying "a variable represents a number" or "a variable assumes a numerical value." These phrases are all intended to suggest the same idea. When numbers and variables are combined using the operations of addition, subtraction, multiplication, and division (and perhaps even square root), **algebraic expressions** are formed. Because we frequently use the letter "x" as a variable, we do not use $\times$ to denote multiplication. We either use the symbol $\cdot$ to denote multiplication or we delete the multiplication sign completely, writing the product of two variables or the product of a number and a variable simply as xy or $3x$.

Consider some more examples in light of these ideas.

EXAMPLE 1 $x + 3$.

"$x + 3$" is an algebraic expression and x is a variable; for any particular value which x assumes, we add 3 to that number to compute $x + 3$. For example, if x assumes the value 5, then $x + 3$ assumes the value $5 + 3$ or 8. Notice that $x + 3$ itself is also a variable; it represents many different numbers. "3," of course, is not a variable; it is a fixed number.

EXAMPLE 2 $x^2 + 3x$.

The symbol x^2 (read "x squared" or "the square of x") means $x \cdot x$. We talk about this kind of notation in the next chapter. As in Example 1, $x^2 + 3x$ is an algebraic expression, and x is a variable that can assume any value. In this example x appears more than once. It is important to observe that when x represents a number, it represents that same number each time it appears in the expression. Thus, when x assumes the value 4, the expression $x^2 + 3x$ assumes the value $4^2 + 3 \cdot 4$ or $16 + 12 = 28$.

EXAMPLE 3 $x^2 + 3y$.

In this example x and y are intended to denote different variables; each can assume any numerical value, and the value assumed by one variable is independent of the value assumed by the other. If we look at this expression where x assumes the value 4, the expression becomes $4^2 + 3y$; y can assume any value. If y assumes the value 3 when x assumes the value 4, the expression becomes $4^2 + 3 \cdot 3$ or 25. If y assumes the value 4 when x assumes the value 4, the expression becomes $4^2 + 3 \cdot 4$ or 28. The expression $x^2 + 3y$ assumes a numerical value for any combination of one value of x and one value of y.

EXAMPLE 4 $y \cdot (2 + y)$.

The parentheses indicate that the entire expression $2 + y$ is multiplied by y. For example, if y is 5, then $2 + y$ is 7 and $y \cdot (2 + y)$ is $5 \cdot 7$ or 35. If y is 9, then $y \cdot (2 + y)$ is $9 \cdot 11$ or 99. The expression $y \cdot (2 + y)$

is different from $y \cdot 2 + y$. In the second expression, if y is 5 then $y \cdot 2 + y$ is $10 + 5$ or 15; if y is 9, then $y \cdot 2 + y$ is $18 + 9$ or 27.

In evaluating the expressions in the examples above for different values of the variables, we have used principles that govern the order in which operations are performed. Stated briefly, these principles are as follows.

ORDER OF OPERATIONS

1. If the expression does not contain any parentheses, work from left to right doing all multiplications and divisions. Then return to the left and perform all additions and subtractions.
2. If the expression does contain parentheses, use principle 1 *inside* the parentheses to find a number. Replace the parenthetical expression with the number and proceed by principle 1.

For further work with these principles and for a general discussion of how to use a calculator, see Appendix B.

Exercises 1-B

I. In problems 1 to 10, use the given variable to write an algebraic expression for the result of the operations described.

EXAMPLES: 5 is added to the variable t. $\underline{t + 5 \ (\text{or } 5 + t)}$

The variable P is decreased by 7. $\underline{P - 7}$

1. The variable n is added to 34. _____
2. The variable x is multiplied by 15. _____
3. A number t is multiplied by 5 and 7 is added to the result. _____

4. 7 is added to a number t and the result is multiplied by 5. _____

5. The variable K is increased by 9. _____
6. The square of y is decreased by 1. _____
7. 10 is added to the product of 7 and t. _____
8. The product of K and 5 is subtracted from the square of K. _____

9. x is increased by 2 and the result is squared. _____

10. 1 is added to the variable n and 1 is subtracted from the variable n; then the two results are multiplied together. _____

11. Write an expression for the number of cents in t quarters. _____

12. Write an expression for the number of cents in s dimes. _____

13. Write an expression for the number of cents in t quarters and s dimes. _____

14. Write an expression for the number of inches in k feet. _____

15. Write an expression for the number of quarts in x gallons. _____

II. Evaluate each expression at the specified values of the variable.

EXAMPLES: $x + 5$ when $x = 2$ _7_
                    $x + 5$ when $x = 7$ _12_

1. $3x + 4$ when $x = 2$ _____
                           $x = 5$ _____
                           $x = 0$ _____

2. $3(x + 4)$ when $x = 2$ _____
                             $x = 5$ _____
                             $x = 0$ _____

3. $5(2x + 1)$ when $x = 1$ _____
                              $x = 5$ _____
                              $x = 10$ _____

4. $(x + 1)(x + 2)$ when $x = 1$ _____
                                    $x = 5$ _____
                                    $x = 0$ _____

5. $2x(x + 2)$ when $x = 1$ _____
                              $x = 3$ _____
                              $x = 0$ _____

6. x^2 when $x = 1$ _____
                       $x = 3$ _____
                       $x = 4$ _____

7. 2^3 when $x = 1$ _____
                       $x = 2$ _____
                       $x = 4$ _____

8. $2t^2$ when $t = 3$ _____
                        $t = 2$ _____
                        $t = 5$ _____

9. $(2s)^2$ when $s = 3$ _____
                          $s = 2$ _____
                          $s = 5$ _____

10. $(x + 2)^2$ when $x = 1$ _____
                              $x = 3$ _____
                              $x = 0$ _____

11. $x^2 + 4$ when $x = 1$ _____

$x = 3$ _____

$x = 0$ _____

12. $(x^2 + 1)(2x + 1)$ when $x = 0$ _____

$x = 2$ _____

$x = 4$ _____

13. $(2x + 1)(2x - 1)$ when $x = 1$ _____

$x = 2$ _____

$x = 4$ _____

14. $4x^2 - 1$ when $x = 1$ _____

$x = 2$ _____

$x = 4$ _____

III. Evaluate the following when the variables represent the numbers given.

EXAMPLE: $x + 2y$; $x = 3$, $y = 5$

Answer: $3 + 2(5) = 13$

1. $3x - y$; $x = 6$, $y = 2$
2. $x(y - 1)$; $x = 3$, $y = 5$
3. $a^2 + 2b$; $a = 2$, $b = 3$
4. $a^2 + 2b$; $a = 2$, $b = 2$
5. $a + b - 2c$; $a = 7$, $b = 4$, $c = 4$
6. $(x - 1)(y + 1)$; $x = 5$, $y = 4$
7. $3 + 2x^2 - y$; $x = 1$, $y = 3$
8. $x^2 - x(y + 1)$; $x = 6$, $y = 1$
9. $y + 2(3 + x)$; $y = 1$, $x = 7$
10. $2(t + s)^2 - 6$; $s = 0$, $t = 5$

IV. Solve the following problems.

1. A car gets 23 miles per gallon on the open road and 17 miles per gallon on in-town driving. The gas tank holds 24 gallons of gas. How far can the car go, under each of the following conditions, on a full tank of gas?

   (a) All the driving is done on the Interstate.

   (b) All the driving is done in town.

   (c) The car uses 9 gallons on a trip on I-95 and the rest on in-town driving.

2. Repeat problem 1 for a car that has a 20-gallon tank and gets 37 miles per gallon on the Interstate (open road) and 26 miles per gallon in town.

3. If in problem 2 the car uses N gallons of gasoline on a trip on I-95 and the rest of the gas on in-town driving, write an expression that shows how far the car can go on a full tank of gas.

4. Write a brief set of directions telling a motorist how to compute the in-town mileage of a car, that is, the distance the car travels on one gallon of gas when the driving is done in town.

1.4 VALUES ASSUMED BY VARIABLES

We have said that a variable stands for the numbers in some set of numbers. We need to look more closely at what numerical values can be assumed by a specific variable. Often the situation that gives rise to the variable provides this information for us. Look again at two of our examples.

1. In the problem on bank charges, we wrote the monthly charges as $0.50 + 0.10N$ where N denotes the number of checks returned that month. What set of numbers can N represent? Clearly, N could not be a fraction or a negative number. Presumably, it could be any one of the natural numbers: 1, 2, 3, ...

2. When we looked at the area of rectangles that have a perimeter of 48 inches, we represented the area as $\ell \cdot (24 - \ell)$ where ℓ is the length of one side of the rectangle. What can be said about ℓ? Clearly, ℓ cannot be a negative number since length is never negative. Also, ℓ must be a number less than 24 since the perimeter is 48 and the rectangle has four sides. In fact, ℓ could represent any number between 0 and 24.

When a variable occurs in an algebraic expression, we assume it represents any number that is sensible in the problem. If the context of the problem puts no restrictions on the numbers the variable can represent, we assume it can represent any number.

Perhaps you are now having some uneasy thoughts similar to these:

> *"I thought algebra was about there being one value for x and that the student was always supposed to find it, somehow. Now you are saying x can be any number. Is this different from what I was taught?"*

A couple more examples should help to clarify the situation. An **equation** is a statement that two algebraic expressions are equal. For example, $3x + 2 = 10 - x$ is an equation; it tells us that the two algebraic expressions $3x + 2$ and $10 - x$ are equal. Although x may represent any number in these algebraic expressions, not every number replacing x will make a true statement. If $x = 1$, then $3x + 2 \neq 10 - x$ because $5 \neq 9$; if $x = 5$, then $3x + 2 \neq 10 - x$ because $17 \neq 5$. In fact, the only value for x that makes the statement true is $x = 2$; $3(2) + 2 = 10 - 2$ because $8 = 8$. When some of us studied high school algebra, a variable in an equation was called an **unknown.** The directions to the exercises commonly said, "Find the unknown." In our language this is the same as asking for a given equation, "Of all the numbers a variable can represent, which ones make the equation a true statement." We sometimes see that short-ened simply to, "Solve the equation."

Some equations contain more than one variable. We have the example where bank charges B can be computed in terms of the number N of checks returned; $B = 0.50 + 0.10N$ is an equation with two variables, and we have said N can represent any number in the set $\{0, 1, 2, 3, 4, \ldots\}$. If the equation is to be a true statement, then the values that B assumes are determined by the values that N assumes. If N is 1, then B is $0.60. If N is 5, then B is $1. A solution for this equation is made up of two numbers, one replacing N and one replacing B. We have a solution if B is 0.60 and N is 1; we have a different solution if B is $1 and N is 5. There are a great many solutions to this equation. (Give four more.)

It is important to keep in mind the difference between an alge-braic expression and an equation.

An algebraic expression is the result of combining numbers and variables with the usual operations of arithmetic.

Examples of Algebraic Expressions:

$$x \qquad 100 \qquad 3x^2 + 2 \qquad xy - 4 \qquad xz - y(3xz)(y + 2x)$$

An equation is a statement of equality between two algebraic ex-pressions.

Examples of Equations:

$$x + 1 = 100 \qquad y = x + 5 \qquad p = 2\ell + 2w \qquad d = rt$$

Variables that occur in algebraic expressions can generally take on all numerical values that make sense in the problem. The question we ask about equations is different; we usually want to know which of the many values that a variable or variables can assume make the equation a true statement. Such values are called **solutions** to an equation. In later chapters we develop techniques for finding solu-tions to particular types of equations, but that is not our immediate concern. In the next section we first find ways to simplify algebraic expressions.

Exercises 1-C

I. Indicate which of the following are algebraic expressions and which are equations.

1. $3(2x - 1)(x + 1)$
2. $3x - x$
3. $15x = 2y + 1$
4. $y - z = (x + 1)(2x)$
5. $x^2 + 3x - 10$

II. For each of the following equations, decide whether the proposed values give solutions to the equation.

1. $3x + 6 = 0$
   $x = 0$ $x = 2$ $x = 6$

2. $10 - 2x = 0$
   $x = 0$ $x = 4$ $x = 5$

3. $(x - 3)(x - 2) = 0$
   $x = 4$ $x = 2$ $x = 3$

4. $t^3 - 2t = 4$ (t^3 means $t \cdot t \cdot t$)
   $t = 0$ $t = 2$ $t = 3$

5. $y = 4x - 8$
   $x = 3$ and $y = 4$ $x = 2$ and $y = 0$ $x = 5$ and $y = 5$

6. $r(s + 2) = r + 3$
   $r = 2$ and $s = 0$ $r = 4$ and $s = 2$ $r = 1$ and $s = 2$

III. Each of the following tables shows an example of a relationship between the "first number" and the "second number." For each, first describe the relationship in words and add some additional number pairs to the table. Then write an equation that shows how we can express the second number y in terms of the first number x.

	First Number	Second Number
EXAMPLE:		
	0	0
	1	2
	2	4
	3	6
	4	8

In words: "The second number is twice the first."

As an equation: $y = 2x$.

1.

First Number	Second Number
0	1
1	3
2	5
3	7
4	9

3.

First Number	Second Number
0	2
1	7
2	12
3	17
4	22

2.

First Number	Second Number
1	2
2	4
3	8
4	16
5	32

4.

First Number	Second Number
0	0
1	1
2	4
3	9
4	16

	First	Second
5.	Number	Number
	1	2
	2	6
	3	12
	4	20
	5	30

IV. Solve the following problems by experimentation and reasoning. You may find that you guess several times before you hit the right answer to the problem.

1. A drive-in movie charges $4 for the car and driver and $1.50 for each passenger. How many people can go to the movie in one car for $10?

2. Ron wants to fence in a square section of his yard as an exercise area for his dog. He got a good deal on 80 feet of fencing. How big a square can he fence with this? How many square feet will the dog have to play in?

3. A cashier went to the bank and got $56 worth of change in nickels and quarters. The coins were in rolls and each roll contained 40 coins. She got twice as many rolls of nickels as quarters. How many rolls of each coin did she get? (*Hint:* First find the value of a roll of nickels and a roll of quarters.)

4. The Miller family is making a 300-mile car trip to attend cousin Anne's wedding. They need to arrive by 4:30 P.M. If they expect to average 50 miles per hour and spend one hour eating lunch, what time should they leave home.

5. Leon purchased a shirt for $9.60 on a one-fourth-off sale. What was the price of the shirt before it was marked down?

REVIEW PROBLEMS

1. Evaluate each expression at the specified values of the variables.
   (a) $25t + 1$ at $t = 2$, $t = 1$, $t = 0$
   (b) $4x(x + 2)$ at $x = 5$, $x = 2$, $x = 0$
   (c) $6 + y(3 + y)$ at $y = 10$, $y = 2$, $y = 0$

2. Evaluate the following when the variables represent the numbers given.
   (a) $5x + 3y + 2xy$, $x = 2$ and $y = 3$
   (b) $(2x + y)(x + 2y)$, $x = 5$ and $y = 1$

3. For each of the following equations, tell which of the proposed values give solutions to the equation.
   (a) $2x - 8 = 0$; $x = 2$ $x = 0$ $x = 4$ $x = 8$
   (b) $(x - 5)(x - 4) = 0$; $x = 1$ $x = 4$ $x = 5$ $x = 20$
   (c) $y^2 + 2y = 3$; $y = 0$ $y = 1$ $y = 2$ $y = 3$

4. Describe the relationship between the first number and the second number in the following tables. Write an equation that shows how the second number y can be expressed in terms of the first number x.

(a)

First (x)	Second (y)
0	0
1	4
2	8
3	12
4	16

(b)

First (x)	Second (y)
0	3
1	5
2	7
3	9
4	11

5. A caterer provides light refreshments for parties at a cost of $50 base fee plus $3 per person. What is the cost of having the caterer for a party of 42 persons?

6. A city parks department wants to spread Turfbuilder on a series of playing fields. The total length of the fields is 1500 feet and the width is 500 feet. How many bags of Turfbuilder, each covering 5000 square feet, will the city need?

CHAPTER 2

EXPONENTS AND ARITHMETIC PROPERTIES

2.1   MULTIPLYING SEVERAL NUMBERS

When you first learned to multiply, you probably memorized a multiplication table that included the products of any two 1-digit numbers. Later you learned procedures for finding the product of any two numbers. Although you may not have given it much thought at the time, all these rules you learned for multiplication concerned the multiplication of *two* numbers. However, there are occasions when a product contains more than two numbers and we need to understand what this means.

Consider, for example, that the owner of a shoe store wants to compute the number of shoe boxes stacked on a storage shelf. If he counts that the space filled is 11 boxes high, 4 boxes deep, and 15 boxes wide, then he could reason that there are $11 \cdot 4$ or 44 boxes in each column and 15 columns, making a total of $44 \cdot 15$ or 660 boxes. Alternatively, he might reason that there are 11 layers of boxes and each layer contains $4 \cdot 15$ or 60 boxes, so altogether there are $11 \cdot 60$ or 660 boxes.

In the first case of the example, the computation is $(11 \cdot 4) \cdot 15$ with the product of 11 and 4, inside the parentheses, computed first and then multiplied by 15. In the second case, the computation is $11 \cdot (4 \cdot 15)$, meaning 11 times the product of 4 and 15. Notice that each time we take a product we are multiplying exactly two numbers together. The way that we group the numbers does not affect the product; this fact is called the **associative property of multiplication.** In our example, associativity says that $(11 \cdot 4) \cdot 15 = 11 \cdot (4 \cdot 15)$. In general, associativity says that, for any numbers n, m, and t, $(n \cdot m) \cdot t = n \cdot (m \cdot t)$. Because of associativity, we frequently write the product of three numbers as $11 \cdot 4 \cdot 15$, omitting the parentheses because we know it is irrelevant where they are placed. Similarly, we may write the product of more than three numbers without parentheses.*

There are times when we multiply a number by itself several times. For example, $10 \cdot 10 \cdot 10$ or $5 \cdot 5 \cdot 5 \cdot 5$. Exponents are helpful in abbreviating these products. In Chapter 1 you solved a problem concerning the growth of bacteria; in that problem exponents would simplify the expressions involved. Exponents also provide a simple form for the numbers that arise in problems concerning population growth, radioactive decay, compound interest, and other topics.

When we write 2^5, we mean $2 \cdot 2 \cdot 2 \cdot 2 \cdot 2$.

---

* Although multiplication and addition are associative operations, not all operations are associative. For example, subtraction is not associative: $(8 - 4) - 2$ does not equal $8 - (4 - 2)$. Thus, parentheses are necessary with subtraction to make the meaning clear.

When we write 3^4, we mean $3 \cdot 3 \cdot 3 \cdot 3$.

When we write 11^2, we mean $11 \cdot 11$.

More generally, if b is any number and n is a natural number, 1, 2, 3, 4, . . . , then b^n is the product of n numbers each of which is equal to b. We agree that b^1 means b.

In the expression b^n, the number b is called the **base** and the number n is called the **exponent.** Thus, a natural number used as an exponent indicates how many times the base appears in the product.

The symbol b^n is read

$$\text{``}b \text{ to the } n\text{''}$$

or

$$\text{``}b \text{ to the } n\text{th''}$$

or

$$\text{``the } n\text{th power of } b\text{''}$$

For example, b^4 is usually read "b to the fourth" or "the fourth power of b," and b^5 is usually read "b to the fifth." However, b^2 is usually read "b squared" (since $b \cdot b$ is the area of a square with side length b) and b^3 is usually read "b cubed" (since $b \cdot b \cdot b$ is the volume of a cube with all sides of length b).

Here is another problem where the computation in the solution can be simplified by using exponents.

PROBLEM FOR EXAMINATION AND DISCUSSION

Suppose you receive a "chain letter" from a friend. The letter contains a list of 5 names and instructs you to send $1 to the first name on the list and remove that name from the list. Move each name up one position and add your name at the bottom. Then copy the letter exactly and mail a copy to 5 new friends. The letter promises that you will receive thousands of dollars in the mail. If the chain works perfectly, how much money will you receive?

In this problem, which you can work on in class or with friends, there is no immediate method we can apply to reach a solution. First, you must analyze what is going on in the problem; you may want to act out the situation until you see exactly what is happening. Such procedures require time and often several sheets of paper, but they usually give a solution. As we proceed, we can develop efficient methods of solving certain types of problems. However, there are always important and interesting problems for which we do not have efficient methods. These we will solve by experimentation and analysis and perseverance. A few problems in the next exercise set are of this type.

Exercises 2-A

I. In each of these equations, find the value of the exponent. (A calculator may be helpful for these problems.)

1. $2^x = 16$
2. $(2^t)2 = 32$
3. $10^x = 1{,}000{,}000{,}000$ *count the zeros*
4. $2^w = 64$
5. $2^x \cdot 3^x = 36$

6. $11^n = 161{,}051$
7. $5^t = 625$
8. $3^x \cdot 3^x = 3^6$ *729*
9. $4^y = 64$
10. $2^y \cdot 5^y = 10{,}000$

II. In each of these equations, find the base.

1. $x^3 = 8$
2. $x^3 = 64$
3. $y^8 = 256$

4. $x^4 = 2401$
5. $u^6 = 729$
6. $5z^3 = 5000$

III. Solve the following problems.

1. (a) Does $(5 + 3)^2 = 5^2 + 3^2$?
   (b) Can you find numbers a and b so that $(a + b)^2 = a^2 + b^2$?
2. A certain single-celled animal splits into two every 30 seconds. If there are 10 of these animals in a tank at 12 noon, how many will be in the tank 1 minute later? 5 minutes later? 1 hour later? x hours later?
3. A culture contains 30 cells. Each cell splits into 2 cells once every half-hour.
   (a) How many cells are there after 3 hours? 5 hours? 10 hours? n hours?
   (b) In how many hours will the culture have 480 cells?
4. The yearly comsumption of energy for industrial purposes has been doubling every 10 years. If E denotes the amount of energy consumed in 1980, and if this increase continues as in the past, what will be the amount of energy consumed in 1990? In 2000? In 2500?

2.2 RULES FOR COMPUTING WITH EXPONENTS

If we consider carefully what exponents mean, we can find ways of simplifying the computations that involve them. Consider first the product of two exponential expressions that have the same base, for example, $3^2 \cdot 3^4$. This represents the number $(3 \cdot 3) \cdot (3 \cdot 3 \cdot 3 \cdot 3)$—the product of six 3s or 3^6. Thus, $3^2 \cdot 3^4 = 3^6$. Here are more examples:

$$2^3 \cdot 2^5 = (2 \cdot 2 \cdot 2) \cdot (2 \cdot 2 \cdot 2 \cdot 2 \cdot 2) = 2^{3+5} = 2^8$$

$$x \cdot x^4 = x \cdot (x \cdot x \cdot x \cdot x) = x^{1+4} = x^5$$

Since the exponents on the left side simply indicate the number of times the base occurs in the factors of the product, the sum of the exponents counts the number of times the base occurs in the product. More generally we can state the observation in rule form in this way.

RULE 1 FOR EXPONENTS

Finding the product of powers of the same base.

$$b^n \cdot b^m = b^{n+m}$$

In this rule we are assuming now that b is any number and that n and m are natural numbers. It is important to remember that this rule applies only to products of terms that have the same base. We can write $t^3 \cdot t^2 = t^5$ or $(2x)^3 \cdot (2x)^5 = (2x)^8$, but the rule says nothing at all about simplifying $3^2 \cdot 2^4$ or $x^2 \cdot y^3$ or $(4x)^2 \cdot (3x)^4$.

There are times when we need to read Rule 1 from right to left; namely, we want to recognize that $b^{n+m} = b^n \cdot b^m$. Before students used calculators, they were taught to use this fact to simplify computations. For example,

$$3^4 = 3^2 \cdot 3^2 = 9 \cdot 9 = 81$$

$$3^6 = 3^4 \cdot 3^2 = 81 \cdot 9 = 729$$

Even with calculators, it is sometimes important to break apart powers in order to simplify expressions.

$$x^6 = x^2 \cdot x^4$$

$$x^{11} = x^3 \cdot x^8$$

$$(3x^2)^3 = (3x^2)(3x^2)^2$$

When we write an expression as a product, we say we have **factored** it. In Chapter 8, we develop factoring skills and consider problems where factoring is important.

Now consider the expression $(4x)^2$. We know that $(4x)^2 = (4x)(4x)$ or $4 \cdot x \cdot 4 \cdot x$. If we interchange the two middle terms, we get $(4x)^2 = 4 \cdot 4 \cdot x \cdot x = 4^2x^2$. The fact that we can change the order of factors and not change the product is called the **commutative property of multiplication:** $a \cdot b = b \cdot a$ for any numbers a and b.* The associa-

---

* Multiplication and addition are commutative but subtraction is not. For example, $6 - 4$ does not equal $4 - 6$.

tive property and the commutative property simplify computation a great deal for us. Consider another example.

$$(3x)^4 = (3x)(3x)(3x)(3x)$$
$$= 3 \cdot x \cdot 3 \cdot x \cdot 3 \cdot x \cdot 3 \cdot x$$
$$= 3 \cdot 3 \cdot 3 \cdot 3 \cdot x \cdot x \cdot x \cdot x$$
$$= 3^4 \cdot x^4$$

From these two examples, $(4x)^2 = 4^2x^2$ and $(3x)^4 = 3^4x^4$, a pattern emerges that we can state as Rule 2 for exponents.

RULE 2 FOR EXPONENTS

Finding a power of a product.

$$(a \cdot b)^n = a^n \cdot b^n$$

This rule is being stated here for any numbers a and b and for a natural number n. It permits us to simplify directly any expression that has the form of a product to some power; for example,

$$(2t)^5 = 2^5 t^5 = 32 t^5$$
$$(xy)^4 = x^4 y^4$$

Rule 2 for exponents says $(2x)^4 = 2^4x^4 = 16x^4$. Sometimes this rule is used in the opposite direction, namely, to write $16x^4$ as $2^4x^4 = (2x)^4$. Here is another example where Rule 2 is read from right to left: $27y^3 = 3^3y^3 = (3y)^3$.

Some expressions that involve exponents can be simplified using both Rules 1 and 2. The following examples illustrate how to combine these rules to simplify efficiently.

EXAMPLE 1   Simplify $(3x)^2 \cdot x^5$.

$$(3x)^2 \cdot x^5 = 3^2 \cdot x^2 \cdot x^5 \quad\quad \text{(Rule 2)}$$
$$= 3^2 \cdot x^7 \quad\quad\quad\quad \text{(Rule 1)}$$
$$= 9x^7$$

EXAMPLE 2   Simplify $(5x)^3 \cdot (2x)^4$.

$$(5x)^3 \cdot (2x)^4 = 5^3 \cdot x^3 \cdot 2^4 \cdot x^4 \quad\quad \text{(Rule 2 used twice)}$$
$$= 5^3 \cdot 2^4 \cdot x^3 \cdot x^4 \quad\quad \text{(Commutative property}$$
$$\text{of multiplication),}$$
$$= 5^3 \cdot 2^4 \cdot x^7 \quad\quad\quad\quad \text{(Rule 1)}$$

$$= 125 \cdot 16 \cdot x^7 \qquad \text{(Computation)}$$
$$= 2000 x^7$$

There is one additional observation that enables us to simplify expressions containing exponents. We want to consider expressions in which the base is itself an exponential expression, for example, $(3^4)^2$ or $(t^3)^5$. Writing the meaning of these expressions suggests a pattern.

$$
\begin{aligned}
(3^4)^2 &= 3^4 \cdot 3^4 \\
&= 3^{4+4} \qquad \text{(Rule 1)} \\
&= 3^{2 \cdot 4} \\
&= 3^8
\end{aligned}
$$

$$
\begin{aligned}
(t^3)^5 &= t^3 \cdot t^3 \cdot t^3 \cdot t^3 \cdot t^3 \\
&= t^{3+3+3+3+3} \qquad \text{(Rule 1)} \\
&= t^{5 \cdot 3} \\
&= t^{15}
\end{aligned}
$$

These are examples of a third rule we can use for computing exponents.

RULE 3 FOR EXPONENTS

Finding a power of a power of a base.

$$(b^n)^m = b^{n \cdot m}$$

Here again at this time b represents any number, and n and m are natural numbers. Rule 3 permits us to write directly $(3^4)^2 = 3^{4 \cdot 2} = 3^8$ and $(t^3)^5 = t^{3 \cdot 5} = t^{15}$. By using all three rules, many expressions can be rewritten so that each variable appears only once in the expression; this is what we mean when we say **simplify** an exponential expression with natural number exponents. Study the following examples carefully.

EXAMPLE 3   Simplify $(3x^3)^2 \cdot (5x^4)^3$.

$$
\begin{aligned}
(3x^3)^2 \cdot (5x^4)^3 &= 3^2 \cdot (x^3)^2 \cdot 5^3 \cdot (x^4)^3 & \text{(Rule 2, twice)} \\
&= 3^2 \cdot x^6 \cdot 5^3 \cdot x^{12} & \text{(Rule 3, twice)} \\
&= 3^2 \cdot 5^3 \cdot x^6 \cdot x^{12} & \text{(Commutative property} \\
& & \text{of multiplication)} \\
&= 3^2 \cdot 5^3 \cdot x^{18} & \text{(Rule 1)} \\
&= 9 \cdot 125 \cdot x^{18} & \text{(Computation)} \\
&= 1125 x^{18}
\end{aligned}
$$

EXAMPLE 4   Simplify $3(2x^3)^5 \cdot x^8$.

$$
\begin{aligned}
3(2x^3)^5 \cdot x^8 &= 3 \cdot 2^5 \cdot (x^3)^5 \cdot x^8 && \text{(Rule 2)}\\
&= 3 \cdot 2^5 \cdot x^{15} \cdot x^8 && \text{(Rule 3)}\\
&= 3 \cdot 2^5 \cdot x^{23} && \text{(Rule 1)}\\
&= 3 \cdot 32 \cdot x^{23}\\
&= 96x^{23}
\end{aligned}
$$

But what, you may be asking, if I forget the rules for computing with exponents? Not all is lost. What you really need to remember is what the exponent means. Computation with exponents is possible without using the rules; however, you may find the computation long and inefficient. Consider Example 1; let's look at its simplification using only the definition of exponent and the associative and commutative properties of multiplication, rather than the rules of exponents that we have observed.

$$
\begin{aligned}
(3x^3)^2 \cdot (5x^4)^3 &= (3x^3)(3x^3)(5x^4)(5x^4)(5x^4)\\[6pt]
&= 3 \cdot x \cdot x \cdot x \cdot 3 \cdot x \cdot x \cdot x \cdot 5 \cdot x \cdot x \cdot x\\
&\quad\; \cdot x \cdot 5 \cdot x \cdot x \cdot x \cdot x\\[6pt]
&= 3 \cdot 3 \cdot 5 \cdot 5 \cdot 5 \cdot x \cdot x \cdot x \cdot x \cdot x \cdot x \cdot x \cdot x \cdot x \cdot x \cdot x \cdot x\\
&\quad\; \cdot x \cdot x \cdot x \cdot x \cdot x \cdot x\\[6pt]
&= 9 \cdot 125 \cdot x^{18}\\[6pt]
&= 1125x^{18}
\end{aligned}
$$

You can see that the rules for computing with exponents permit us to compute efficiently and probably with fewer mistakes. With enough practice, they will become familiar tools in your computation bag.

The laws of exponents provide methods for simplifying expressions involving multiplication; they do not help at all with addition. In fact, terms like $2x^2 + x^5$ cannot be simplified any further. Sometimes a student may want to say that $2x^2 + x^5$ should equal $2x^7$. However,

$$
2x^2 + x^5 = 2 \cdot x \cdot x + x \cdot x \cdot x \cdot x \cdot x
$$

and

$$
2x^7 = 2 \cdot x \cdot x \cdot x \cdot x \cdot x \cdot x \cdot x
$$

If you replace x by any nonzero number, you will see these are different expressions. For example, if $x = 1$, then $2x^2 + x^5 = 3$ and $2x^7 = 2$. Even though we cannot combine terms that involve different powers of a variable or variables, we can combine terms where the powers are the same. For example, we can simplify $2x^5 + x^5$. This is the topic of our next section.

Exercises 2-B

I. Simplify each of the following, using exponents. Each base should appear only once in an expression.

1. $2 \cdot 2 \cdot 2 \cdot 2 \cdot 3 \cdot 3 \cdot 5$
2. $7 \cdot 8 \cdot 7 \cdot 8 \cdot 7 \cdot 8 \cdot 7 \cdot 8$
3. $10 \cdot 10 \cdot 10 \cdot 10 \cdot 10 \cdot 10 \cdot 10 \cdot 10$
4. $(2 \cdot 2 \cdot 2 \cdot 2 \cdot 2 \cdot 2 \cdot 2) \cdot (2 \cdot 2 \cdot 2 \cdot 2)$
5. $3 \cdot 3^3 \cdot 3^5$
6. $2^3 \cdot 3^2 \cdot 5 \cdot 3^2 \cdot 2$
7. $3^3 \cdot 7 \cdot 3^2 \cdot 7^3$
8. $13^3 \cdot 3^{13} \cdot 13^2$
9. $2^5 \cdot 2^5 \cdot 2^5 \cdot 2^5$
10. $(5^3)^2 \cdot 5^2$ *mult. exp. when raising to a power*

II. Write each of the following as x^2 times another expression. For example, $3x^5 = x^2 \cdot 3x^3$.

1. $4x^3$
2. $10x^2y^2$
3. $141x^5y$
4. $3x^2 \cdot (xy)^2$
5. x^5t^5
6. x^2

III. Simplify each of the following expressions. Each base should appear only once in the expression.

1. $(2x)^3$
2. $x^3 \cdot x^5 \cdot x^7$
3. $(2x^2)(3x^5)$
4. $4x^2 \cdot (4x)^2$
5. $x^5 \cdot 5x$
6. $(5x)^2 \cdot (3x)^4$
7. $x^4 \cdot (2x^2)^3$
8. $7t^5 \cdot 5t^7$
9. $(5y^2)(2y^5)$
10. $(2t^3)^2$
11. $(3z^2)^4 \cdot z^3$
12. $(2xy)^3$
13. $(2x^2y)^4$
14. $(x^2y^3)^2(2xy)^3$
15. $(3x^3)^2(2x^3)^2$
16. $3t^2(2xt)^3$
17. $(5t^2x^3)^2(4x^3y)^4$
18. $[2(2x^2)^3]^4$
19. $x^5(2x)(4x^2)$
20. $[x(x^2)(x^3)]^4$
21. $(a^2)^3(2ab)^2$
22. $[(3a^2)^3]^2$
23. $(2x \cdot x^2)^3$
24. $[3(3t^2)]^2$

2.3   THE DISTRIBUTIVE PROPERTY

Think about this situation.

Two women stop at the corner cafe for coffee and donuts before work. Each has a 45¢ cup of coffee and a 30¢ donut. How much is the bill?

There are at least two ways to regard this problem. The first is to reason that each woman spends 0.45 + 0.30 and there are two women so the bill is 2(0.45 + 0.30). A second is to observe that the women are charged for two cups of coffee and two donuts so the bill is 2(0.45) + 2(0.30). Both are correct since 2(0.45 + 0.30) = 2(0.45) + 2(0.30).

The property of arithmetic that says $a(b + c) = a \cdot b + a \cdot c$ for any numbers a, b, and c is called the **distributive property**; more accurately, we say that multiplication distributes over addition in our arithmetic.* This property together with commutativity and associativity, which we have already discussed, guides our computation with numbers and variables.

Figure 2.1

A second illustration of the distributive property can be found in geometry. In Appendix A, we show that the area of a rectangle can be computed by taking its length times its height. A property of area is that, if a region is broken into nonoverlapping parts, the area of the whole region is the sum of the areas of the parts. Thus, in Figure 2.1, the area of the large rectangle, $a \cdot (b + c)$, equals the sum of the areas of the two smaller rectangles, $a \cdot b + a \cdot c$.

Notice that using commutativity we can write the equation $a \cdot (b + c) = a \cdot b + a \cdot c$ in the form $(b + c) \cdot a = b \cdot a + c \cdot a$. The following examples use both of these forms.

EXAMPLE 1
$$x(3x + 2) = x \cdot 3x + x \cdot 2$$
$$= 3 \cdot x \cdot x + 2 \cdot x$$
$$= 3x^2 + 2x$$

EXAMPLE 2
$$x + 2(x^2 + 7) = x + 2x^2 + 14$$

EXAMPLE 3
$$(2y - 1)y = 2y^2 - y$$

EXAMPLE 4
$$(2x + 3)(x + 4) = 2x(x + 4) + 3(x + 4) \quad \text{(Distributive property)}$$
$$= 2x^2 + 8x + 3x + 12 \quad \text{(Distributive property—again!)}$$
$$= 2x^2 + 11x + 12$$

---

* Multiplication also distributes over subtraction: $a(b - c) = a \cdot b - a \cdot c$. For example, $2(15 - 8) = 2 \cdot 15 - 2 \cdot 8$.

EXAMPLE 5
$$(t + 3)(t + 3) = t(t + 3) + 3(t + 3) \quad \text{(Distributive property)}$$
$$= t^2 + 3t + 3t + 9 \quad \text{(Distributive property)}$$
$$= t^2 + 6t + 9$$

An important application of the distributive property is in simplifying algebraic expressions. Notice that in Example 5 we have written $3t + 3t$ as $6t$. Actually, when we do this, we are using the distributive property in the form $b \cdot a + c \cdot a = (b + c) \cdot a$. Study these examples.

$$3t + 3t = (3 + 3) \cdot t = 6t$$

$$3x + 5x = (3 + 5) \cdot x = 8x$$

$$7y^2 + 2y^2 = (7 + 2) \cdot y^2 = 9y^2$$

$$rs^2 + 3rs^2 = (1 + 3) \cdot rs^2 = 4rs^2$$

We sometimes refer to this type of simplification as "combining like terms." Notice that we have said nothing about simplifying $3t + 5t^2$. The distributive property does not help with this and we usually leave the expression as $3t + 5t^2$.

There are times when we informally use the distributive property in computations without recognizing that we have used the property. For example, how does a shopper compute the price of 7 cans of soup at 41¢ each. She can reason $7(0.40) = 2.80$ plus another 0.01 on each can, or \$2.87. In doing this, she has used the fact that

$$7(0.41) = 7(0.40) + 7(0.01)$$

Exercises 2-C

I. Simplify by multiplying and collecting like terms. Use the distributive law where appropriate.

1. $7(x + 4)$
2. $x(2x + 3)$
3. $(k + 2)(k + 7)$
4. $(x + 2)^2$
5. $(t + 3)(2t + 5)$
6. $5 + 9(2x + 3)$
7. $7x + 7(2x + 3)$
8. $9x[1 + 3(2x + 1)]$
9. $3 + 2(x + 5)(x + 1)$
10. $4x[2 + 3(2x + 1)]$
11. $(2t + 3)^2$
12. $z^2(z^2 + 1)$

13. $x(1 + 3x + 2x^2)$
14. $5x(x^2) + 2x^3$
15. $(x^2 + x)^2$
16. $x(3x^2 + x^3) - 2x^2$
17. $(2x^2)^2$
18. $x^2y + xy^2 + 2xy$
19. $x(y + z) + y(x + z)$
20. $2t(t^2 + 1) + 3t(1 + t)$
21. $a(b + c) + b(a + c) + c(a + b)$
22. $a^2x^3[x + 3x(x^2 + x)]$
23. $3y^3[2 + 3y(y + 1)^2]$
24. $[(2x^2 + 1)^2 + x^2]^2$

II. Solve the following problems.

1. Each month the Hendricks pay $385 on their mortgage and $132 on their car. How much do they pay on these two large purchases in a year?

2. A rectangular swimming pool is twice as long as it is wide. The distance around the edge of the pool is 180 feet. What are the dimensions of the pool?

Suppose you wanted to get a canvas cover that would exactly cover this pool in the winter. How many square feet of canvas would you need?

3. Jim's car gets 19 miles per gallon on the open road. Gas costs $1.29 per gallon. How much will it cost him for gas on a 703-mile trip on the Interstate?

4. Donuts cost 19¢ apiece and Danish cost 35¢ apiece. Mrs. Smith bought 48 pastries, mixed donuts and Danish, and the total cost was $13.60. How many of each type did she buy?

5. Joe wants to fence in a rectangular garden beside his house. The house wall will be one side of the garden, and the other three sides will be fenced. He has 60 feet of fence (and will use it all). What will be the area of the garden if he makes it square (see Figure 2.2)?

What will be the area if he makes it twice as long as it is wide?

What is the largest garden he can make with this fence?

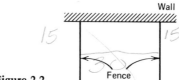

Figure 2.2

III. These problems give more practice computing with exponents. Simplify each expression. The answers are given here so that you can find any error immediately.

1. $(3x^2)^2 \cdot 5x^3$ Answer: $45x^7$

2. $(3x + 2)^2(x^4)$ Answer: $9x^6 + 12x^5 + 4x^4$

3. $x^3y(xy)(2y^2)(y^2 + 1)$ Answer: $2x^4y^6 + 2x^4y^4$

4. $(x^2t + 1)^2 + 3x^3(xt + xt^2)$ Answer: $4x^4t^2 + 2x^2t + 3x^4t + 1$

5. $(3x^2 + 5)^2 + (x + 1)^2$ Answer: $9x^4 + 31x^2 + 2x + 26$

6. $(t^3 + t^2)^2 \cdot (2t^4)^2$ Answer: $4t^{14} + 8t^{13} + 4t^{12}$

7. $[(1 + x)^2 + 1]^2$ Answer: $x^4 + 4x^3 + 8x^2 + 8x + 4$

8. $[5 + 3(2x + 2)]^2$ Answer: $36x^2 + 132x + 121$

9. $[2x^2 + (2x)^2]^2$ Answer: $36x^4$

10. $[5 + (2 + 3x)]^2$ Answer: $9x^2 + 42x + 49$

11. $[(x + 1)(x + 2)]^2$ Answer: $x^4 + 6x^3 + 13x^2 + 12x + 4$

REVIEW PROBLEMS

1. Find the value of x.
   (a) $3^x = 81$
   (b) $x^4 = 625$

2. Simplify each of the following expressions. Each base should appear only once in the expression.
   (a) $3x^2(3x)^2$
   (c) $(4x)^2(4x)^3$
   (b) $(2t^3)^4$
   (d) $3(3x)^2x^5$

3. Simplify by multiplying and collecting like terms.
   (a) $2x(x + 3x^2) - x^2$
   (c) $(2x + 5)^2$
   (b) $(10y - 1)y$
   (d) $[x^2 + (4x)^2]^2$

4. Herb and Sue Miller have invested $4500 in an investment account that doubles in value every 6 years. What is the value of their investment after 6 years? 12 years? 24 years?

5. Sarah is a regional manager for her firm. Her travel allowance is $0.19 per mile plus $65 per diem to cover board and room. How much money will she be reimbursed for five days of travel when she averaged 200 miles a day?

CHAPTER 3

LINEAR EQUATIONS AND INEQUALITIES IN ONE VARIABLE

3.1   COMPARING ALGEBRAIC EXPRESSIONS

Suppose you choose a number and I choose a number. We each write our numbers down and then compare them. We will find that exactly one of the following statements is true.

Our numbers are identical.

Or your number is larger than mine.

Or your number is smaller than mine.

This simple observation reflects the principle of trichotomy in arithmetic.

PRINCIPLE OF TRICHOTOMY

If a and b are any numbers, then exactly one of the following statements is true:

a is less than b (Symbolized $a < b$)
a is equal to b (Symbolized $a = b$)
a is greater than b (Symbolized $a > b$)

[handwritten: numbers are Either equal or one is bigger than the other]

Notice that $a > b$ (a is greater than b) and $b < a$ (b is less than a) have the same meaning. Check that the following statements are true.

$$2 < 7 \qquad 5 + 8 > 0 \qquad 7 - 3 < 5 \qquad 14 - 6 = 8 \qquad 2^3 < 3^2$$

$$4^2 = 2^4 \qquad 5^2 < 2^5 \qquad 6 + 8 = 8 + 6$$

Now consider two algebraic expressions involving the variable x, for example,

$$3x + 2 \qquad \text{and} \qquad 2x + 5$$

In these expressions, x can represent any number. The relationship between the two expressions depends on the value of x. More particularly,

When $x = 1$,   then $3x + 2 < 2x + 5$
because $(3 \cdot 1 + 2) < (2 \cdot 1 + 5)$

When $x = 3$,   then $3x + 2 = 2x + 5$
because $(3 \cdot 3 + 2) = (2 \cdot 3 + 5)$

When $x = 5$,   then $3x + 2 > 2 + 5$
because $(3 \cdot 5 + 2) > (2 \cdot 5 + 5)$

As we know from Chapter 1, a statement that two expressions are equal is called an **equation,** and any value of the variable for which the equation is true is called a **solution** of the equation. Thus, in our example, 3 is a solution to $3x + 2 = 2x + 5$.

A statement that one algebraic expression is greater than, or less than, another expression is called an **inequality.** Any value of the variable for which an inequality is true is called a **solution of the inequality.** In the example above, the number 1 is a solution of the inequality $3x + 2 < 2x + 5$; 0 and 2 are also solutions of this inequality as you can see by computing the values of both expressions when $x = 0$ and again when $x = 2$. Similarly, in our example, 5 is a solution to the inequality $3x + 2 > 2x + 5$; check to see that 7 and 19 are also solutions of this inequality, but that 2 is not a solution.

Our purpose in this chapter is to develop techniques for finding the solutions to certain types of equations and inequalities; in later chapters, we extend these skills to other kinds of equations and inequalities. But first we need to make a few observations.

Not every equation has a solution. For example, there is no number that x can represent to make $x + 2 = x + 3$; adding 2 to a number always gives a different sum than adding 3 to the same number. Some equations have many solutions: every number is a solution to $2(x + 1) = 2x + 2$, for example. We saw in the inequalities relating $3x + 2$ and $2x + 5$ that inequalities can have many solutions. In general, inequalities do have more than one solution. However, some inequalities have no solutions; for example $x + 3 < x + 1$ does not have a solution. (Try it and see!)

Sometimes two equations will have exactly the same solutions. When this happens, we say the equations are **equivalent.** Here are six equivalent equations; each has one solution, the number 4.

$$x = 4 \qquad x + 3 = 7 \qquad x - 2 = 2 \qquad 2x = 8$$

$$4(x + 3) = 28 \qquad 3x - 2 = 2x + 2$$

We also say that two inequalities are equivalent when they have exactly the same solutions. We are now ready to turn our attention to finding the solutions of equations and inequalities.

3.2. SOLVING LINEAR EQUATIONS IN ONE VARIABLE

The equations we have just looked at are of a special type. They have only one variable and the highest exponent on the variable is 1; each term involving the variable is of the form "some number times the variable." Such equations are said to be **linear** in one variable.

Here are some linear equations.

$$3x - 7 = 0$$

$$4(x - 1) = x + 5$$

$$12 - x = 6$$

Here are some equations that are not linear.

$$x^2 + x = -6$$

$$\frac{3}{x} + 2x = 0$$

$$x(x + 1) = 15$$

Our technique for solving a linear equation is usually to replace it with an equivalent simpler one, then to replace the new equation with an equivalent one that is simpler than it, and so on, until we arrive at an equation so simple that its solution is obvious. The question, then, is how to find a simpler equation equivalent to a given equation.

Consider again the equation $3x + 2 = 2x + 5$. If for a given value for x, the number $3x + 2$ is the same as the number $2x + 5$, then (subtracting 2 from both expressions) the number $(3x + 2) - 2$ is the same as the number $(2x + 5) - 2$; that is, the number $3x$ equals the number $2x + 3$. Hence the equation $3x = 2x + 3$ is equivalent to our original equation $3x + 2 = 2x + 5$. But we don't need to stop here. If $3x$ is the same number as $2x + 3$ for some value of x, then $3x - 2x$ is the same as $(2x + 3) - 2x$. Thus, we have another equivalent equation: $3x - 2x = (2x + 3) - 2x$ or $x = 3$. The solution to this last equation is obvious; it is true only when x assumes the value 3.

We can summarize the discussion above by listing the successive equivalent equations.

$$3x + 2 = 2x + 5$$

$$3x = 2x + 3 \qquad \text{(Subtract 2 from both sides)}$$

$$x = 3 \qquad \text{(Subtract } 2x \text{ from both sides)}$$

The point in this example is that, if we have two expressions that are equal for some value of the variable and we subtract the same quantity from each of the expressions, we end up with two new expressions that are equal for the same value of the variable. And, of course, this is not only true for subtraction. If we add the same quantity to two expressions that are equal for the same value of the variable, or if we multiply (or divide) two expressions that are equal by the same nonzero number, we end up with two new expressions that are equal. Thus, any of the following procedures can be trusted to convert a linear equation to an equivalent equation.

- Add the same expression to both sides of the equation.
- Subtract the same expression from both sides of the equation.
- Multiply both sides of the equation by the same nonzero number.
- Divide both sides of the equation by the same nonzero number.

tools for linear equations

These procedures are our tools for solving linear equations. If an equation has a solution, there is more than one sequence of procedures that can be applied to yield the solution. A rule of thumb is to try to collect all terms containing the variable on one side of the equation (it doesn't matter which side) and all terms free of the variable on the other side.

Here are some additional examples. Keep in mind that the goal is to find a value for x that makes the given equation true.

EXAMPLE 1 Solve $7x - 8 = 5x + 8$.

$$7x - 8 = 5x + 8$$

$$7x = 5x + 16 \qquad \text{(Add 8 to both sides)}$$

$$2x = 16 \qquad \text{(Subtract } 5x \text{ from both sides)}$$

$$x = 8 \qquad \text{(Divide both sides by 2)}$$

The equations all have one solution, $x = 8$.

EXAMPLE 2 Solve $3(2x + 2) = 4(x + 3)$.

$$3(2x + 2) = 4(x + 3)$$

$$6x + 6 = 4x + 12 \qquad \text{(Distributive property, both sides)}$$

$$2x + 6 = 12 \qquad \text{(Subtract } 4x \text{ from both sides)}$$

$$2x = 6 \qquad \text{(Subtract 6 from both sides)}$$

$$x = 3 \qquad \text{(Divide both sides by 2)}$$

The equations all have the same solution, $x = 3$.

EXAMPLE 3 Solve $1 + 3(2 - x) = x - 1$.

$$1 + 3(2 - x) = x - 1$$

$$1 + 6 - 3x = x - 1 \qquad \text{(Distributive property, left side)}$$

$$7 - 3x = x - 1 \qquad \text{(Simplify left-hand side)}$$

$$7 = 4x - 1 \qquad \text{(Add } 3x \text{ to both sides)}$$

$$8 = 4x \qquad \text{(Add 1 to both sides)}$$

$$2 = x \qquad \text{(Divide both sides by 4)}$$

The equations all have one solution, $x = 2$.

EXAMPLE 4 Solve $2(3x + 2) = 6x + 4$.

$$2(3x + 2) = 6x + 4$$

$$6x + 4 = 6x + 4 \qquad \text{(Distributive property, left side)}$$

The second equation is true for all values of x; hence the original equation has all numbers for its solutions. If we had failed to observe that all numbers are solutions to equation two, we might have continued the computation this way:

$$6x + 4 = 6x + 4$$

$$4 = 4 \qquad \text{(Subtract } 6x \text{ from both sides)}$$

Again we get an equation that is always true, and we conclude that all numbers are solutions.

EXAMPLE 5 Solve $2(3x + 2) = 6x + 3$.

$$2(3x + 2) = 6x + 3$$

$$6x + 4 = 6x + 3 \qquad \text{(Distributive property, left side)}$$

The second equation is not true for any value of x. The equations have *no* solution. Again, if we did not see that $6x + 4 = 6x + 3$ has no solution, we could continue the computation.

$$6x + 4 = 6x + 3$$

$$4 = 3 \qquad \text{(Subtract } 6x \text{ from both sides)}$$

This equation is clearly never true, and we can conclude that there are no solutions.

Sometimes an equation involves more than one variable. When we have found an equivalent equation that expresses the value of one variable in terms of the other variables, we say we have solved the equation for the one variable.

EXAMPLE 6 Solve $a = 2c - b$ for c.

$$a = 2c - b$$

$$a + b = 2c \qquad \text{(Add } b \text{ to both sides)}$$

$$\frac{a + b}{2} = c \qquad \text{(Divide both sides by 2)}$$

EXAMPLE 7   Solve $2(3t + 5s) = 3(4t + 2s)$ for s.

$$2(3t + 5s) = 3(4t + 2s)$$

$6t + 10s = 12t + 6s$	(Distributive property)
$10s = 6t + 6s$	(Subtract $6t$ from both sides)
$4s = 6t$	(Subtract $6s$ from both sides)
$s = \dfrac{6t}{4}$	(Divide both sides by 4)

Now it is time for *you* to solve some equations; the only way to really learn how to solve them is through practice. Sometimes a student is able to follow examples that have been worked out, but has difficulty solving equations alone. The feeling that "I don't know what to do first" is not uncommon. It may well be that there is more than one thing that can sensibly be done "first." There is no best way to solve every equation. The important thing is to remember that you want to end up with a value for the variable. The equation is the starting point; the value for the variable (the solution!) is the goal. The properties of arithmetic and the procedures for finding equivalent equations are the rules of the game.

Exercises 3-A

I. Solve the following equations.

1. $5x = 15$
2. $3x = 27$
3. $8 = 2t$
4. $12 = z + 7$
5. $3x + 2 = 11$
6. $7y - 2 = 19$
7. $3 + 4x = 27$
8. $8x - 2 = 46$
9. $4x - 12 = 48$
10. $7y + 2 = 51$
11. $9 + 6t = 45$
12. $3(x + 2) = 18$
13. $5(x - 2) = 25$
14. $3(2t + 1) = 45$
15. $2y + 5 = y + 8$
16. $4t + 5 = 9 + 3t$
17. $3x + 4 = 20 - x$
18. $5t - 2 = 3t + 2$
19. $2(2x - 1) = 3x + 17$
20. $4x + 3 = 5x + 1$
21. $2(x + 4) = 3(x - 1)$
22. $6(x + 1) = 3(x + 19)$
23. $2x + 1 = 5(10 - x)$
24. $5(x - 7) = 2(2x - 6)$
25. $4x + 6 = 4x$
26. $2x + 4 = 2(x + 2)$

II. Solve each of the following equations for the variable indicated.

1. $P = 2\ell + 2w$; solve for ℓ.
2. $P = 2\ell + 2w$; solve for w.

   3. $c = m \cdot d + b$; solve for b.
   4. $2a + 3b = 2(a + b + c)$; solve for b.
   5. $3w - 5z = 3z - 5w$; solve for w.
   6. $C = 2\pi r$; solve for r.

III. Write an equation that describes the situation in each of the following problems.

   1. A child pays 11¢ for each of several pencils plus 15¢ for one eraser. If he spends $1.14, how many pencils should he have?

   2. The Athletic Boosters buy pompoms for 65¢ each and sell them for $1. If their profits are $189, how many did they sell?

   3. A rectangular-shaped garden is 2 feet longer than it is wide, and the total length of fence enclosing it is 40 feet. What are the dimensions?

   4. A car rental agency charges $19.95 per day plus 20¢ per mile. How many miles were traveled if the charges for one day were $123.95?

   5. A car travels at 55 miles per hour. How much time is required for the car to go 165 miles?

   6. A fruit drink is 40 percent fruit juice. How much water should be added to 20 gallons of the fruit drink to dilute it to a drink that is 25 percent fruit juice? (If you are not comfortable with percent, consult Appendix C.)

   7. A teenager leaves her home by car at 60 miles per hour. Two hours later her anxious parents leave along the same road at 80 miles per hour. How long is it before the parents overtake the teenager?

3.3   SOLVING LINEAR INEQUALITIES IN ONE VARIABLE

Recall that an inequality is an assertion that one algebraic expression is less than or greater than another expression. For example, $3x + 2 < 2x + 5$ is an inequality. As with equations, the usual question is what numerical values can the variable represent to make a true statement. In our example, the statement is true for $x = 0$ or $x = 1$ or $x = 2$ but is not true for $x = 3$ or $x = 4$ or $x = 5$. Numbers that make the inequality a true statement are called its solutions. The vocabulary for inequalities is analogous to that for equations. For example, the two inequalities that follow are called linear because the variables appear with exponent 1 but not to any higher exponent, and the variables do not appear as divisors.

$2x + 3 > x + 7$ has 5, 6, 7, and 8 as solutions (check them!). In fact, any number greater than 4 is a solution; try some others.

$t + 3 < t$ has no solutions; explain why.

We need techniques for finding the solutions of inequalities. Our methods are analogous to those for equations. Two inequalities are called equivalent if they have exactly the same solutions. The idea is to replace an inequality with an equivalent but simpler inequality and repeat until the solutions become obvious. If one expression is less than another for some value of the variable and we add (or subtract) the same quantity to each, the result is two new expressions with the first again less than the second for the same values of the variable. Thus, we can add the same quantity to both sides of an inequality, and the result is an equivalent inequality. Or we can subtract the same quantity from both sides, and the result will be an equivalent inequality. We can make the same statements for multiplication and division as long as we are restricting ourselves to positive numbers. When we introduce negative numbers, we will investigate the effect of multiplying and dividing the expressions in an inequality by negative numbers, and we will see that the story is quite different. For now, we have these procedures for changing one linear inequality into an equivalent inequality.

- Add the same expression to both sides.
- Subtract the same expression from both sides.
- Multiply both sides by the same *positive* number.
- Divide both sides by the same *positive* number.

See how these procedures are used in each of the following examples involving linear inequalities.

EXAMPLE 1   Solve $6x - 7 > 13 + 2x$.

$$6x - 7 > 13 + 2x$$

$$6x > 20 + 2x \qquad \text{(Add 7 to both sides)}$$

$$4x > 20 \qquad \text{(Subtract } 2x \text{ from both sides)}$$

$$x > 5 \qquad \text{(Divide both sides by 4)}$$

All numbers greater than 5 are solutions to this inequality. To assure yourself that this is the solution, pick several numbers greater than 5 and test them in the original inequality. Also try 5 and some numbers less than 5.

EXAMPLE 2   Solve $6x > 2x$.

$$6x > 2x$$

$$4x > 0 \qquad \text{(Subtract } 2x \text{ from both sides)}$$

$$x > 0 \qquad \text{(Divide both sides by 4)}$$

All numbers greater than 0 are solutions to this inequality.

EXAMPLE 3   Solve $5(2x - 9) < 3(x + 6)$.

$$5(2x - 9) < 3(x + 6)$$

$$10x - 45 < 3x + 18 \qquad \text{(Distributive property)}$$

$$10x < 3x + 63 \qquad \text{(Add 45 to both sides)}$$

$$7x < 63 \qquad \text{(Subtract } 2x \text{ from both sides)}$$

$$x < 9 \qquad \text{(Divide both sides by 7)}$$

All numbers less than 9 are solutions to this inequality.

A WORD ABOUT NOTATION

Consider the following problem.

Marsha has $250 to spend on four new tires for her car. What price can Marsha afford for an individual tire?

If we let x represent the price of a tire, then Marsha can afford the tires if $4x < 250$ or if $4x = 250$. Rather than consider these separately, it is convenient to combine them. To do this we introduce a new symbol, $\leq$, and write $4x \leq 250$. We read this as "$4x$ is less than or equal to 250." Marsha can afford the tires provided that $4x \leq 250$ or $x \leq 62.50$.

The symbol "$\leq$" can be used to relate any two algebraic expressions. Consider $3x \leq 2x + 4$. The solutions to $3x \leq 2x + 4$ consist of the solutions to $3x < 2x + 4$ and the solutions to $3x = 2x + 4$. In this example, we might solve the inequality and the equation separately as follows.

$$3x < 2x + 4$$

$$x < 4 \qquad \text{(Subtract } 2x \text{ from both sides)}$$

$$3x = 2x + 4$$

$$x = 4 \qquad \text{(Subtract } 2x \text{ from both sides)}$$

Thus, we see that the solutions to $3x \leq 2x + 4$ consist of the number 4 and all numbers less than 4. A more efficient way to produce this solution is to combine the solving of the inequality and the equation in this way.

$$3x \leq 2x + 4$$

$$x \leq 4 \qquad \text{(Subtract } 2x \text{ from both sides)}$$

We can do this because our procedures for solving equations and inequalities are the same when we work with positive numbers.

The symbol $\geq$ is used to mean "either greater than or equal to." We compute with it in a way analogous to the way in which we compute with $\leq$.

EXAMPLE 4   Solve $4(x - 3) \leq 3(x + 2)$.

$$4(x - 3) \leq 3(x + 2)$$

$$4x - 12 \leq 3x + 6 \qquad \text{(Distributive property)}$$

$$4x \leq 3x + 18 \qquad \text{(Add 12 to both sides)}$$

$$x \leq 18 \qquad \text{(Subtract } 3x \text{ from both sides)}$$

The solutions to this inequality are 18 and all numbers less than 18.

Exercises 3-B

Solve the following inequalities.

1. $2x < x + 5$	**13.** $4x + 3 > 5x + 1$
2. $4x - 3 > 3x + 1$	**14.** $5(x - 10) > 2x + 25$
3. $5(x + 2) \leq 3(x + 10)$	**15.** $11x + 10 < 10x + 11$
4. $4x \geq 3x$	**16.** $4(x + 3) > 4x + 12$
5. $6x - 9 < 3x + 12$	**17.** $4(x + 3) \geq 4x + 12$
6. $4x + 2 \geq 5x - 7$	**18.** $2(x + 1) > 5(x - 11)$
7. $3x + 4 > 14 - 2x$	**19.** $6(x + 7) \leq 7(x + 6)$
8. $2(5x + 3) < 3(10 + 2x)$	**20.** $2x + 3(7 + x) > 4(x + 6)$
9. $6x - 30 \geq 0$	**21.** $2(x + 5) > 3x + 2$
10. $6(x + 1) < 3(13 + x)$	**22.** $11(3x - 10) \leq 8(x + 30)$
11. $3x + 4 \geq 3(x + 1)$	**23.** $5(x - 9) \geq 2(x + 24)$
12. $3x + 4 \leq 3(x + 1)$	**24.** $543x - 687 > 851 - 226x$

3.4   GRAPHING SOLUTIONS TO EQUATIONS AND INEQUALITIES

In the last section we developed algebraic techniques for solving linear equations and inequalities in one variable. The solutions were sets of numbers. Now we want to give geometric descriptions of these numerical sets. To get started we take a line, assign the number 0 to one point, and then assign each natural number (in order) to a point on the line using equal spacing. We cannot sketch every point on a line because the line extends indefinitely in both directions and we cannot write every whole number, but we can represent the idea as we have in Figure 3.1.

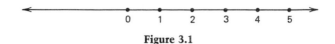

Figure 3.1

Although we have limited our attention so far to the numbers, 0, 1, 2, 3, 4, . . . , we know there are many numbers between these; there are also numbers less than 0. These numbers are described in Chapter 4.

The solutions to the inequality $x \leq 4$ consist of 4 and all numbers less than 4. Thus, the solutions can be drawn as in Figure 3.2.

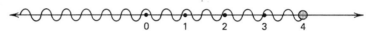

Figure 3.2

The picture of the solutions of an equation or inequality is called the *graph* of the equation or inequality. The graph of $x < 4$ consists of all numbers less than 4 (but not 4). Figure 3.3 illustrates the solutions to this inequality.

Figure 3.3

Notice that we have used an empty circle to indicate when 4 is not a solution and a filled circle to indicate when 4 is a solution.

Here are some additional examples of graphs of inequalities and equations.

EXAMPLE 1   Graph the solutions to $x \geq 2$.

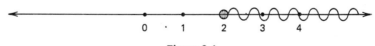

Figure 3.4

EXAMPLE 2   Graph the solutions to $2x = x + 3$.

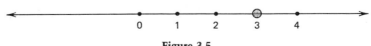

Figure 3.5

EXAMPLE 3   Graph the solutions to $2x < x + 3$.

Figure 3.6

EXAMPLE 4   Graph the numbers that are solutions to *both* $x \geq 3$ and $7 > x$.

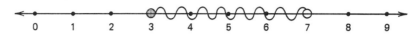

Figure 3.7

Exercises 3-C

I. Graph the solutions to the equations given in problems 1 to 5 of Exercises 3-A, Part I. Graph the solutions to the inequalities given in problems 1 to 5 of Exercises 3-B.

II. For each of the following, graph the numbers that are solutions to *both* of the given inequalities.

1. $x \leq 9$ and $x > 2$.   4. $x < 8$ and $3 \leq x$.
2. $x \leq 9$ and $x \leq 7$.   5. $x > 3$ and $x \leq 1$.
3. $x > 5$ and $x > 1$.

III. Solve the following problems.

 EXAMPLE: A donut shop sells 24 different types of donuts at different prices ranging from 17¢ to 29¢ each. Write inequalities describing D, the cost of a dozen donuts at this shop.

   12 of the least expensive donuts
   cost $12 \cdot 17$¢ or \$2.04   so $D \geq 2.04$
   12 of the most expensive donuts
   cost $12 \cdot 29$¢ or \$3.48   so $D \leq 3.48$

1. A small hotel has 12 small bedrooms and 15 large ones. Small rooms can each accommodate 1 or 2 people. Large rooms can each accommodate up to 4 people. Write inequalities to describe N, the number of people staying in the hotel when each room has at least one occupant. Graph the solutions.

2. Every workday Todd Jones packs sandwich lunches for his family. Depending on whether there are after-school activities that day, each of the four children carries one or two sandwiches. Todd's wife takes 1 sandwich, and Todd takes 2 or 3. Write inequalities to describe x, the number of sandwiches Todd packs on a given day. Graph the solutions.

3. Todd's wife Jan does the weekly food shopping. As her personal index of inflation, Jan keeps track of the average price per bag of

groceries each week. In recent months her best price for 7 bags of groceries was \$42; last week she was appalled to find that her bill for 7 bags was \$66.50, an all-time high. Using x as the cost per bag, write the inequalities describing the cost of groceries during recent months.

4. A bank charges a service charge of 75¢ per month plus 10¢ for each check processed. How many checks were processed in a month for which the service charge was \$4.95?

5. Mr. and Mrs. Smith have a joint account at the bank described in the previous problem. In April the Smith's service charge was \$4.25, and Mrs. Smith wrote 4 times as many checks as Mr. Smith. How many checks did each write?

6. Steak costs \$1.60 more per pound than ground chuck. Mrs. Jones bought 3 pounds of steak and 4 pounds of ground chuck and her total bill was \$18.03. What is the price per pound of ground chuck? Of steak?

7. Truxall Plumbing charges a \$25 one-time charge for coming out to a house for one-half hour or less, and \$20 for each hour (or part of an hour) beyond that time. The Moore's had extensive remodeling done. The plumber came out on Monday and Tuesday. The total bill for the job was \$225. How many hours did the plumber spend at the Moore's?

IV. These problems provide more practice in solving linear equations and inequalities. Solve each of the following. The answers are given here so that you can can find any error immediately.

1. $9x - 2 = 5(x + 2)$ — Answer: $x = 3$
2. $5(x - 2) = 4(x + 10)$ — Answer: $x = 50$
3. $3(x - 4) = 2(9 - x)$ — Answer: $x = 6$
4. $x - 37 = 37 - x$ — Answer: $x = 37$
5. $3x + 4 = 3(x + 4)$ — Answer: no solution
6. $400 - 7x = 8x + 175$ — Answer: $x = 15$
7. $2x + 3(x - 3) = 2(x + 10) + 4$ — Answer: $x = 11$
8. $11x + 1 = 6(x + 6)$ — Answer: $x = 7$
9. $x + 3(x + 4) = 4(x + 2) + 4$ — Answer: x can be any number
10. $5(x + 4) = 4(x + 5)$ — Answer: $x = 0$
11. $5(x - 2) > 5$ — Answer: $x > 3$
12. $3x - 24 \le 16 - x$ — Answer: $x \le 10$
13. $7(x - 1) < 4x + 14$ — Answer: $x < 7$
14. $3(x - 2) \le 14 - x$ — Answer: $x \le 5$
15. $3(x + 1) \ge 3 - 2x$ — Answer: $x \ge 0$
16. $7x + 6 < 2x + 101$ — Answer: $x < 19$
17. $3x + 2 \ge 3(x + 2)$ — Answer: no solution
18. $5(6x + 1) > 13(x + 3)$ — Answer: $x > 2$
19. $5(x + 2) - 2x > 3x + 7$ — Answer: x can be any number
20. $15x + 136 \le 3x + 700$ — Answer: $x \le 47$

REVIEW PROBLEMS

1. Solve the following equations.

   (a) $7x = 63$ (d) $4(x + 3) = 60$

   (b) $19 = 12 + y$ (e) $7x - 2 = 3(x + 2)$

   (c) $5t - 2 = 8$ (f) $7x + 8 = 8x + 7$

2. Solve each of the following for the variable indicated.

   (a) $3a + 2b = a + b + c$; solve for b.

   (b) $d = rt$; solve for t.

3. Solve each of the inequalities and graph the solution.

   (a) $3x + 2 > 8$

   (b) $4 - x < 2 + x$

4. A college charges part-time students tuition of $330 for the first three credits and $95 for each additional credit. Jane's tuition this term is $710; how many credits is she taking?

5. Tom's trip from his Columbus office to his Chicago office requires twice as much driving time at 25 miles per hour as flying time averaging 320 miles per hour. If the total number of miles traveled is 370 miles, how long does he spend driving? How long does he spend flying?

he drives to the airport + to his office from the airport in Chic.
he does not drive to Chic.

CHAPTER 4

EXPANDING THE NUMBER SYSTEM: INTEGERS

4.1   USING NEGATIVE NUMBERS

In the first three chapters of this book, we have mostly used natural numbers and 0. Our reason for doing this has been to help you focus on and understand the basic algebraic ideas of variable, equation, inequality, and solution in the context of familiar numbers that you use every day. Now that you have mastered these ideas we want to expand our number system to include negative numbers (and later fractions) and to examine how negative numbers are used in computations, equations, and inequalities.

Two common uses of negative numbers in our day-to-day lives are found in measuring very low temperatures and in the recording of deficits by bookkeepers. We consider first the use of negative numbers in each of these contexts.

A thermometer is in many ways similar to the number line constructed in the last chapter (see Figure 4.1). Numbers are equally spaced along the side of a thin tube of mercury. On a Celsius (metric system) thermometer, 0 marks the height of the mercury when it is just cold enough for water to freeze. As mercury gets warmer, it expands and rises in the tube, and we can read the temperature as the number of spaces (or degrees) above 0 that are filled by the mercury. When the weather becomes colder than the freezing point of water, the mercury contracts; it does not read as high as the 0 point. In order to describe the temperature, we extend our equally spaced marks below zero and count how many marks below zero the top of the mercury is. On the thermometer in Figure 4.1, the mercury is 2 marks (or degrees) below zero; another way of reading this is to say that the mercury reaches −2 on the thermometer. If the temperature goes up by 3 degrees from −2, it will be −2 + 3 or 1 degree (above 0). If it goes down 5 degrees from −2 it will be −2 − 5 or −7 degrees.

The Friendly Bank allows each checking account customer a "ready reserve," that is, a certain amount of money which can be borrowed from the bank when the customer writes a check for more money than is on deposit in the account. When "ready reserve" funds are used by a customer, the balance in the account is recorded as a negative number. Study the Bank Statement in Figure 4.2 to see how this works. Notice that, on February 9, the customer's check was larger than the balance available; therefore, after the check was cashed, the balance was negative. Notice what happens to negative balances as money is added to the account by deposits and subtracted from the account by check cashing.

Negative numbers are useful in many applications of mathematics. Generally speaking, negative numbers are useful in any situa-

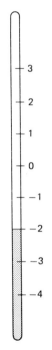

Figure 4.1

Bank Statement			
Date	Transaction	Amount	Account Balance
2/5	Beginning Balance		$235.59
2/7	Cash check #324	$125.00	110.59
2/7	Cash check #326	89.43	21.16
2/9	Cash check #327	100.00	−78.84
2/10	Deposit	50.00	−28.84
2/14	Deposit	50.00	21.16
2/16	Cash check #325	45.99	−24.83
2/18	Cash check #328	10.00	−34.83

Figure 4.2

tion in which numbers are used to express quantities that can vary above and below some threshold (such as the freezing point of water or the state of having no funds in a bank account). In scientific applications negative numbers are used in other ways as well; for example, scientists might think of the forces pushing an object in one direction as positive forces and those pushing in the opposite direction as negative. In Chapter 6, we will see how positive and negative numbers are used to indicate the orientations of lines in a plane. For the present we turn our attention to a description of negative numbers and rules for computing with them.

4.2   ADDING INTEGERS

The collection of numbers 1, 2, 3, 4, . . . , together with the negative numbers −1, −2, −3, −4, and the number 0 is called the set of **integers**. Our examples have suggested situations where integers are useful; these examples also can suggest what we mean by the sum of two integers. Of course, we already know what it means to add two positive integers, or a positive integer and 0. What we need to investigate is the sum of a positive and a negative integer, and the sum of two negative integers.

We can use the Friendly Bank example in Figure 4.2 as a guide. Study the following table and see if the results are sensible to you.

Expression	Interpretation	Result
10 + 5	A balance of $10 followed by a deposit of $5	15
10 + (−5)	A balance of $10 followed by a withdrawal of $5	5
5 + (−10)	A balance of $5 followed by a withdrawal of $10	−5
−5 + 10	A balance of $−5 followed by a deposit of $10	5
−10 + 5	A balance of $−10 followed by a deposit of $5	−5
−5 + (−10)	A balance of $−5 followed by a withdrawal of $10	−15
5 + (−5)	A balance of $5 followed by a withdrawal of $5	0

Using the Bank Statement as a model, find the sums of the following integers. The answers are given so that you can check your reasoning.

(a) 102 + 50   (e) −40 + (−25)
(b) 102 + (−50)   (f) 45 + (−55)
(c) −15 + 10   (g) −10 + 50
(d) 35 + (−35)   (h) 20 + 30

Answers

(a) 152   (b) 52   (c) −5   (d) 0
(e) −65   (f) −10   (g) 40   (h) 50

Our thermometer example gives us a second way to visualize the sum of two integers. Study this summary.

Expression	Interpretation	Result
10 + 5	Temperature of 10° followed by an increase of 5°	15
10 + (−5)	Temperature of 10° followed by a decrease of 5°	5
5 + (−10)	Temperature of 5° followed by a decrease of 10°	−5
−5 + 10	Temperature of −5° followed by an increase of 10°	5
−10 + 5	Temperature of −10° followed by an increase of 5°	−5
−5 + (−10)	Temperature of −5° followed by a decrease of 10°	−15
−10 + 10	Temperature of −10° followed by an increase of 10°	0

Now interpret the exercises you have just completed, thinking of the thermometer example to be sure you get the same answers.

It is somewhat difficult to summarize in words the procedure for adding two integers when one, or both, of them is negative. One way of approaching this description is to observe that every negative integer corresponds to a positive integer; for example, −5 corresponds to 5, −17 corresponds to 17, −10 corresponds to 10. As we have seen in the examples, $5 + (-5) = 0$ and $(-10) + 10 = 0$. In fact, for any integer n, $n + (-n) = 0$. For this reason, we say that $-n$ is the *opposite* of n and that n is the opposite of $-n$. Thus, -5 is the opposite of 5 and 10 is the opposite of -10.

Using the idea of opposites, we can describe the procedure for adding two negative integers in this way.

To find the sum of two negative integers add their opposites as whole numbers and then take the opposite of that sum.

EXAMPLE 1 Find $(-3) + (-8)$.

Add the opposites: $\qquad\qquad\qquad\qquad$ $3 + 8 = 11$
Take the opposite of the sum: $\qquad$ -11
Thus, $(-3) + (-8) = -11$.

The procedure for finding the sum of a positive integer and a negative integer is more difficult to describe in words. It helps to have the phrase, "sign of a number." When the symbol $+$ or $-$ is used to indicate that a number is positive or negative, the symbol is called the *sign* of the number. When no symbol precedes a number, its sign is $+$.

To find the sum of a positive number and a negative number consider the positive number and the opposite of the negative number; subtract the smaller from the larger; attach the sign of whichever original number is farther from 0.

This description actually involves three steps. Even though with practice we do not write all three steps, it is useful to think this way initially.

EXAMPLE 2 Find $-5 + 7$.

Consider the positive number and the opposite of the negative number: $\qquad\qquad\qquad\qquad\qquad\qquad$ 7, 5
Subtract the smaller from the larger: $\qquad$ $7 - 5 = 2$
Attach the sign of the original number that is farther from 0 (7 is farther from 0 than -5 is): $\qquad$ $+2$

Thus, $-5 + 7 = 2$.

EXAMPLE 3 Find $(-8) + 3$.

Consider the positive number and the opposite of the negative number: $\qquad\qquad\qquad\qquad\qquad\qquad$ 3, 8
Subtract the smaller from the larger: $\qquad$ $8 - 3 = 5$

Attach the sign of the original number that is farther
from 0 (−8 is farther from 0 than 3 is);   −5

Thus, (−8) + 3 = −5.

Note that the parentheses in this problem are not necessary; they
serve only to emphasize the number −8.

Return now to the exercises you completed on page 00 and apply
these procedures to some of them.

Exercises 4-A

Compute the following sums.

1. 10 + (−13)
2. (−37) + (−28)
3. (−83) + 79
4. (−205) + (199)
5. 108 + (−113)
6. (−108) + 113
7. (−175) + (−225)
8. (−39) + (58)
9. (−93) + (58)
10. 207 + (−109)
11. [2 + (−2)] + (−3)
12. 2 + [(−2) + (−3)]
13. [(−42) + 37] + (−10)
14. (−42) + [37 + (−10)]
15. (−3) + (−4) + 5 + (−6) + (−7)

4.3   SUBTRACTING INTEGERS

In previous examples we interpreted 10 + (−5) as the "temperature
of 10° followed by a decrease of 5°" and as "a bank balance of $10
followed by a withdrawal of $5." Both of these situations could also
have been represented by 10 − 5. Indeed, adding −5 is the same as
subtracting 5. This observation becomes our guide for subtracting
integers: **subtracting an integer is the same as adding its opposite.**
Here are some examples.

(a) 12 − 5 = 12 + (−5) = 7
(b) 2 − 5 = 2 + (−5) = −3
(c) −1 − 5 = −1 + (−5) = −6
(d) 8 − (−6) = 8 + (6) = 14
(e) −7 − (−6) = −7 + (6) = −1
(f) −2 − (−6) = −2 + (6) = 4

In examples (a), (b), and (c) where a positive integer is being subtracted, you may be able to visualize "12 decreased by 5" or "2
decreased by 5" or "−1 decreased by 5" without an intermediate
step in the computation. However, in (d), (e), and (f) where a negative
integer is being subtracted, it is well to remember that subtracting

an integer is the same as adding its opposite and in these cases to record, at least mentally, the intermediate step.

A word of caution: Computations involving subtraction of negative numbers are often troublesome for students who are inexperienced or "rusty" with the procedures. This may be because the same symbol "$-$" is used to denote that a number is negative and also to denote the operation of subtraction. This double usage probably developed because subtracting a number is the same as adding its opposite.

Recall that it is not always possible to subtract a natural number from another natural number and get a natural number. For example, $8 - 15$ is not a natural number. With the integers, we can always find the difference of any two natural numbers. Indeed, we can find the difference of any two integers. Some of your computing with integers will likely be done on a calculator. Most calculators handle negative numbers very easily once you learn how to use them. The following computations are done for you so that you can check whether your methods of computing by hand and by calculator are giving you correct answers. Appendix B, at the back of the book, tells you how to use the calculator's $\boxed{+/-}$ key for negative numbers.

$$-5 + 8 = 3 \qquad\qquad 15 - (-8 + 12) = 11$$
$$-5 - 8 = -13 \qquad\qquad -7 - (4 - 18) = 7$$
$$-5 - (-8) = 3 \qquad\qquad -3 - (4 - 7) = 0$$
$$-5 + (-3) = -8 \qquad\qquad 4 - 5 - 6 = -7$$
$$-4 - (-6) + (-7) = -5 \qquad (3 - 8) - (3 - 3) = -5$$
$$-3 - (-8 - 6) = 11 \qquad (7 - 5) - (5 - 7) = 4$$

Exercises 4-B

I. Find the value of each of the following expressions.

1. $-6 + 12$

2. $-5 + (-9)$

3. $-6 + 4 + (-3)$

4. $8 + (-17)$

5. $-8 + (-17)$

6. $-6 + (-13) + 32$

7. $5 - 8$

8. $-7 - 12$

9. $-9 - (-3)$

10. $-4 + [2 - (-3)]$

11. $-3 - 8 + 17$

12. $12 - (-8) - (22)$

13. $-4 - (-3) - (-4)$

14. $-4 - [-3 - (-4)]$

15. $5 - (-6 - 8)$

16. $-3 - (-8 + 14)$

17. $(3 - 5) - (5 - 3)$

18. $(-4 - 6) - [6 - (-4)]$

19. $2 - [3 - (4 - 5)]$

20. $-1 - [-2 - (-3 - 4)]$

II. Beginning with February, fill in the missing data; note that some of the data for each month depends on the previous months.

Month	Precipitation This Month	Precipitation This Year	Normal For Year	Difference Between Precipitation This Year and Normal
January	5 inches	5 inches	6 inches	−1 inch
February	4	?	13	−4
March	6	?	19	?
April	?	?	25	−3
May	10	?	33	?

4.4   MULTIPLYING INTEGERS

We want now to describe the product of two integers. The case of positive integers and 0 is not new; we retain what we have always used for products of natural numbers and 0. It is the product of a positive and a negative integer and the product of two negative integers that we still must describe. Our Friendly Bank and our thermometer models can help us visualize the product of a positive and a negative integer but are not much help with the product of two negative integers.

Remember that for natural numbers $3 \cdot 4$ means "three fours" or $4 + 4 + 4$; that is, for natural numbers, multiplication is repeated addition. In the bank model $3 \cdot 4$ can be interpreted as 3 deposits of $4 each. This is the interpretation we want to extend to a positive integer times a negative integer; interpret $3 \cdot (-4)$ to mean 3 withdrawals of $4 each. Thus,

$$3 \cdot (-4) = (-4) + (-4) + (-4) = -12$$
$$2 \cdot (-65) = (-65) + (-65) = -130$$
$$6 \cdot (-3) = (-3) + (-3) + (-3) + (-3) + (-3) + (-3) = -18$$

We want this multiplication to be commutative so we agree that

$$(-4) \cdot 3 = 3 \cdot (-4) = -12$$
$$(-65) \cdot 2 = 2 \cdot (-65) = -130$$
$$(-3) \cdot 6 = 6 \cdot (-3) = -18$$

With these conventions we see then that the product of a positive and a negative integer is always a negative integer.

To describe the product of two negative integers, we acknowledge that we always want 0 times a number to be 0 and we want the distributive property to hold. Then to describe $(-3) \cdot (-4)$ we can reason this way.

$$4 + (-4) = 0 \quad \text{so} \quad (-3) \cdot [4 + (-4)] = 0$$
$$\text{thus} \quad (-3) \cdot 4 + (-3)(-4) = 0$$
$$\text{or} \quad -12 + (-3)(-4) = 0$$

We see that $(-3)(-4)$ must be 12. Similar reasoning shows us that the product of two negative integers is always a positive integer and, in fact, the product equals the product of the opposites of the negative integers.

We might add here a brief word about division. It is not always possible to divide one integer by another and get an integer. We will see later that division by an integer is equivalent to multiplication by a fraction; therefore, the properties for division are determined by the properties for multiplication. A consequence of this is that the result of the computation is negative when a positive and a negative number are multiplied or divided; the result is positive when both numbers are negative (or, of course, when both numbers are positive).

Exercises 4-C

I. Compute each of the following.

1. $5 - (-8)$

2. $-7 - (-8)$

3. $9 - (-6)$

4. $9 + (-6)$

5. $-11 - (10) - 2$

6. $-(-8)(-6)$

7. $-3(2 - (-6))$

8. $-8 - 3(-2 - 5 \cdot 3)$

9. $-3[-1 - (-2 - 3)]$

10. $6 - 3\{4 - [-8 - (-3)]\}$

11. $(-3)(-4 - 5) - (-5)[6 - (15)]$

12. $1 - [(1 - 2) - (-3)(5 - 7)]$

13. $(5 - 3) - (3 - 5)$

14. $(7 - 9) - (9 - 7)$

15. $-[(6 - 9) - 3(5 - 2)] - (-5)$

II. Compute each of the following.

1. $(-2) + (-3)$

2. $(5) + (-7)$

3. $(-5) + 7$

4. $4 + (-3) + (-6)$

5. $2 + (-3 + 1)$

6. $(-3)(5)$

7. $(-3)(-6)$

8. $2 + (-3) \cdot (7)$

9. $-3 + 2[-1 + (-2)]$

10. $-5 + (-3)(6 + 8)$

11. $4 + (-3 + 6) \cdot (-2)$

12. $[-3 + (-2)] \cdot [-4 + (-6)]$

13. $(-3 + 5) \cdot [3 + (-5)]$

14. $[-3 \cdot (-2 + 7) \cdot (-3 + 4)] + (-2)$

15. $(-2)^2$

16. $(-2)^3$

17. $(-2)^4$

18. $(-2)^5$

19. $(-3)^2$

20. $(-3)^3$

21. $(-3)^4$

22. $(-5)^2$

23. $(-5)^3$

24. **(a)** Compute $(-1)^n$ for 15 different values of n.

   (b) Can you write a rule for the value of $(-1)^n$ that works for any natural number n?

4.5   INTEGERS AND VARIABLES

In our discussion of variables in Chapter 1, we said that when a variable occurs in an algebraic expression it can represent any number which makes sense in the problem. At that time we were talking about positive numbers; however, the statement remains true when we expand our number system to include *all* integers. Unless there are indications to the contrary, a variable can assume any integer value. Thus, in an expression such as $2x - 1$, x can now represent any integer and $2x - 1$ can assume many positive and negative values: $3 = 2(2) - 1$, $-5 = 2(-2) - 1$, $-103 = 2(-51) - 1$, and many others. In many problem situations, such as those involving temperature and bank balances, variables often assume negative values; in other problems, such as those where variables represent length or area, the problem context implies that the variable assumes only positive values.

A WORD ON NOTATION

You will frequently see the algebraic expression $-x$. When x assumes a positive value such as 19, $-x$ assumes the negative value, -19. On the other hand, when x is negative $-x$ is positive; for example, if $x = -3$, then $-x = 3$. Therefore, as a consequence of the fact that x can assume both positive and negative values, $-x$ can also be either positive or negative. The symbol $-x$ is commonly read "the opposite of x."

Simplification and evaluation of expressions involving $-x$ are analogous to procedures you have become familiar with in previous chapters. The expression $-x$ can be thought of as (-1) times x. Use the fact that $-x = (-1)(x)$ and apply the principles for simplifying expressions and for computation with integers as you do the following exercises.

Exercises 4-D

I. Evaluate each of the following expressions for the values of x that are given.

EXAMPLE: $x^2 - x$ if $x = 5$ $5^2 - 5 = 20$

$x = 3$ 6

$x = 0$ 0

$x = -3$ 12

$x = -5$ 30

1. $2x^2$ if $x = 2$ _____

$x = 1$ _____

$x = 0$ _____

$x = -1$ _____

$x = -2$ _____

2. $x^3 + x$ if $x = -1$ _____

$x = -2$ _____

$x = 2$ _____

$x = 1$ _____

3. $3x^3 + (-x)$ if $x = 2$ _____

$x = 1$ _____

$x = 0$ _____

$x = -1$ _____

$x = -2$ _____

4. $-3x^2$ if $x = 2$ _____

$x = 1$ _____

$x = -1$ _____

$x = -2$ _____

$x = -3$ _____

5. $4x^3 + 2x^2$ if $x = -2$ _____

$x = -1$ _____

$x = 0$ _____

$x = -3$ _____

6. $(2x)^3 + 2x^3$ if $x = -1$ _____

$x = -2$ _____

$x = -3$ _____

7. $4 - x$ if $x = 10$ _____

$x = 5$ _____

$x = 0$ _____

$x = -5$ _____

$x = -10$ _____

8. $(2 - x)(2 + x)$ if $x = 4$ _____

$x = 2$ _____

$x = 0$ _____

$x = -2$ _____

$x = -4$ _____

II. Simplify.

1. $1 + (-x)^4$
2. $1 + (-x)^3$
3. $(-2x)^5$
4. $(3x^2) \cdot (-x)^5$
5. $(-x^6) + (-x)^6$

6. $(-x)^7 + (-x)^7$
7. $(-3x^2)^3 \cdot (-2x)^4$
8. $[(-5x) \cdot (-3x^2)^3]^3$
9. $-(-3x)^4$
10. $-4x^2(-5x^2 \cdot 2x^3)^3$

III. Use the distributive property to simplify the following.

EXAMPLE: $5x - 3(x^2 + x) = 5x - 3x^2 - 3x$
$$= 2x - 3x^2$$

1. $-3x^2(-2x + 5)$
2. $3x - 2(-2x + x^2)$
3. $(-5x)^3(1 - x^2)$
4. $2 - 3x[x^4 - (-x)^3]$
5. $-5x[-3x^2 + (-7)]$

6. $5 - 3x^2(5 - 3x^2)$
7. $-2x[x^2 - 3(2 - x)]$
8. $4 - x^3[7x^2 + (-2x - x^6)]$
9. $-[5 - 3(4 - x)]$
10. $1 - \{2 - [3 - (4 - x^2)]\}$

4.6 ORDERING THE INTEGERS

In Chapter 3, we observed that for any two natural numbers, *a* and *b*, either $a > b$, $a = b$, or $a < b$. We also placed the natural numbers on a line. We did this by choosing a point to correspond to 0 and by marking off equal spaces to the right of 0. One effect of this procedure was that the numbers became greater as we moved to the right; in fact, for any pair of natural numbers *a* and *b*, $a > b$ could be interpreted as "*a* lies to the right of *b*."

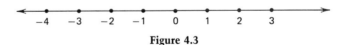

Figure 4.3

We illustrate negative numbers on the number line by marking off equal spaces to the *left* of 0 (see Figure 4.3). For example, -2 is placed two spaces to the left of 0. Notice that the same relationship between inequalities and relative positions of points on the line still applies.

$a > b$ provided *a* lies to the right of *b*

$a < b$ provided *a* lies to the left of *b*

Thus, $-1 > -3$ because -1 lies to the right of -3; $-10 < -5$ because -10 lies to the left of -5.

The solutions of inequalities involving negative numbers can be represented on the number line using the techniques in Section 3.4. For example, the solutions to $x < -4$ are shown on the number line in Figure 4.4.

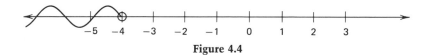

Figure 4.4

On the number line in Figure 4.5, the solutions of $x \geq -2$ are graphed.

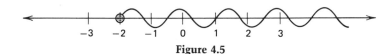

Figure 4.5

Exercises 4-E

I. For each pair of integers, insert the appropriate relation symbol $<$, $=$, or $>$.

1. $-2, 2$	**6.** $5 - 12, 5 - 10$
2. $-4, -5$	**7.** $-8 - 6, -8 - 5$
3. $0, -3$	**8.** $5 - 7, 7 - 5$
4. $-100, -101$	**9.** $-2 \cdot 3, -3 \cdot 2$
5. $0 - 3, 0 - 2$	**10.** $-3 \cdot 4, -3 \cdot 5$

II. Place each group of integers on a number line.

1. $-2, 2, -4, 4$

2. $-9, 6, 0, -1, 3$

3. $-100, -90, 0, 10$

III. Graph the solutions of each of the following inequalities.

1. $x < 5$	**4.** $x \leq -6$
2. $x < -3$	**5.** $x > -10$
3. $x \geq -2$	

4.7   PROPERTIES OF INTEGER ARITHMETIC

We have described the addition, subtraction, and multiplication of integers in a way that makes it clear how to compute with these numbers. However, we have not described carefully the properties of these operations. Before we turn our attention to solving equations and inequalities with integers, it will be helpful to summarize these properties. They provide the "rule book" for simplifying equations and inequalities.

PROPERTIES OF ARITHMETIC

In this table a, b, and c can be any integer.

Properties of Addition

$a + b$ is an integer	
$a + b = b + a$	(Commutative property of addition)
$a + 0 = a$	(0 is the additive identity)
$(a + b) + c = a + (b + c)$	(Associative property of addition)

Properties of Multiplication

$a \cdot b$ is an integer	
$a \cdot b = b \cdot a$	(Commutative property of multiplication)
$a \cdot 1 = a$	(1 is the multiplicative identity)
$a \cdot 0 = 0$	(Multiplicative property of 0)
$(a \cdot b) \cdot c = a \cdot (b \cdot c)$	(Associative property of multiplication)

Meaning of Subtraction

$$a - b = a + (-b)$$

Distributive Properties

$$a \cdot (b + c) = a \cdot b + a \cdot c$$
$$a \cdot (b - c) = a \cdot b - a \cdot c$$

4.8 SOLVING EQUATIONS AND INEQUALITIES INVOLVING INTEGERS

The procedures for solving equations using integers is precisely the same as that described in Chapter 3. For example, consider this equation and solution.

$$-8x - 15 = -5x - 3$$

$-8x = -5x + 12$	(Add 15 to both sides)
$-3x = 12$	(Add $5x$ to both sides)
$x = -4$	(Divide both sides by -3)

The procedures described in Chapter 3, together with the properties of arithmetic for integers, enable us to determine the integer solutions of any linear equation.

In order to solve inequalities involving integers efficiently, however, we need to add to the list of procedures (see Chapter 3, page 39) for changing a linear inequality into an equivalent linear inequality. In particular, we need to define a procedure for finding a simpler inequality that is equivalent to one like $-3x < -18$ or $-5x > 20$.

In order to decide what this procedure should be, examine the effects of multiplying or dividing both sides of an inequality by positive and negative quantities. The following table shows what happens in several examples. Study them carefully.

Sample Inequality	Multiply Both Sides by 2	Multiply Both Sides by −2
$1 < 3$	$2 < 6$	$-2 > -6$
$-2 < 1$	$-4 < 2$	$4 > -2$
$-3 < -1$	$-6 < -2$	$6 > 2$
$-4 > -5$	$-8 > -10$	$8 < 10$
$3 > -4$	$6 > -8$	$-6 < 8$
$-3 < 0$	$-6 < 0$	$6 > 0$

Notice that in each of the examples the direction of the inequality is unchanged when both sides are multiplied by 2. However, when both sides are multiplied by -2, the direction of the inequality is reversed; this happens whenever both sides of an inequality are multiplied by a negative number. The following principle applies.

> If $a < b$, and n is negative,
> then $n \cdot a > n \cdot b$.

We can see that this principle follows from our rules for multiplication of integers. The reasoning is as follows.

$a < b$ means $b - a$ is positive. If n is negative, $n \cdot (b - a)$ is negative, so $n \cdot b - n \cdot a$ is negative, which means that $n \cdot a > n \cdot b$.

This principle leads us to the final procedures we need in solving linear inequalities.

- Multiply both sides by the same *negative* number and *reverse* the direction of the inequality.
- Divide both sides by the same *negative* number and *reverse* the direction of the inequality.

Using these procedures together with the four given on page 39 of Chapter 3, we are able to solve any linear inequality. See how these procedures are applied in solving the following two linear inequalities.

SOLVE

$$-8 - 5x > 12$$
$$-5x > 20 \qquad \text{(Add 8 to both sides)}$$
$$x < -4 \qquad \text{(Divide both sides by } -5 \text{ and change direction)}$$

The solutions to this inequality are all numbers less than -4. The graph of these solutions is given in Figure 4.6.

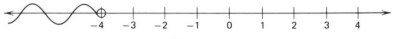

Figure 4.6

SOLVE

$$6 - 2x \geq 2x - 6$$
$$-2x \geq 2x - 12 \qquad \text{(Subtract 6 from both sides)}$$
$$-4x \geq -12 \qquad \text{(Subtract } 2x \text{ from both sides)}$$
$$x \leq 3 \qquad \text{(Divide both sides by } -4 \text{ and change direction)}$$

The solutions to this inequality are the number 3 and all numbers less than 3. The graph of the solutions to the inequality $6 - 2x \geq 2x - 6$ is given in Figure 4.7.

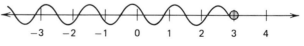

Figure 4.7

Exercises 4-F

I. Solve the following inequalities.

1. $-5 - 3x < 7$ 2. $2 - 4x > x - 8$

3. $1 - x > x - 1$

4. $2(6 - x) > 3 + x$

5. $8 + 7x < 7 + 8x$

6. $-2(2 + x) - 3x \geq 12 - 2x$

7. $3(1 - x) < 10 - x$

8. $4(3 - x) > 3(4 - x)$

9. $-6(-2 - x) < -2(-1 - x)$

10. $-3(x + 4) < 6(x + 4)$

II. Graph the numbers that are solutions to *both* of the given inequalities.

1. $2x > 6; -3x > -12$

2. $-x > 6; -x < 11$

3. $-5x > -10; -3x > 6$

4. $-3x < 0; -4x < 12$

5. $-3x < 6; -x > 5$

III. Solve the following problems.

1. Sue and Joe Johnson pay their gas bills on the budget plan. The gas company set their monthly payments in 1980 at $62 based on their gas usage of the previous year. In the following table, record the difference between the budget payment and the actual cost for each month. In August the gas company computes the actual gas used during the year and then either charges or credits the Johnsons for the difference between the payments made and the actual costs for the year. What was the situation in August 1981?

Month	Budget Payment	Cost for Actual Use	(Budget Payment) − (Actual Cost)
September	$62.00	$ 31.18	
October	62.00	47.15	
November	62.00	78.54	
December	62.00	91.12	
January	62.00	131.53	−69.53
February	62.00	124.52	
March	62.00	101.88	
April	62.00	45.14	
May	62.00	39.75	
June	62.00	33.37	28.63
July	62.00	24.40	
August	62.00	23.91	

2. In Toronto the anticipated high temperature for the day is given in Celsius degrees. An American visitor computes each morning the Fahrenheit measure that corresponds to the Celsius measure in the morning newspaper. She uses the equation $F = 1.8 C + 32$. What temperature can she expect for each of the following Celsius values.

$$C = 0 \quad C = 1 \quad C = -2 \quad C = -20 \quad C = 30$$

REVIEW PROBLEMS

1. Find the value of each of the following expressions.
   (a) $-5 + (-13) + 30$ (c) $-11 - 2(5 + 3)$
   (b) $15 - (3 - 8)$ (d) $(-3)^2 + (-3)^3$

2. Evaluate $4x^2 - x$ for $x = 3$; for $x = -1$.

3. Simplify the following.
   (a) $(-x^2)^3(3x)$
   (b) $4x - 2(-2x + x^2)$

4. Place these integers on a number line: $-3, (-3)^2, (-3)^3, -2, (-2)^2, (-2)^3$.

5. Solve the following inequalities.
   (a) $3 - x < x + 9$
   (b) $-2(x + 4) > 3x + 7$

6. Dow-Jones Industrials closed last Friday at 778.92. The next week the market behaved in this way.

Monday:	Up 2.66
Tuesday:	Down .38
Wednesday:	Down .12
Thursday:	Up 1.40
Friday:	Up 2.06

   Where did the market close this Friday?

7. Charlie Davis is renting space at the State Fair for his Sno-Cone machine. He believes his daily profit can be computed as
   $$P = 0.40N - 50$$
   where N represents the number of cones sold in a day. Find his profit P if $N = 100$; $N = 150$; $N = 300$.

CHAPTER 5

EQUATIONS IN TWO VARIABLES AND THEIR GRAPHS

5.1   PICTURING RELATIONSHIPS AS GRAPHS

Imagine that your neighbor is planning a trip by car to visit a relative who lives 300 miles away. The time it takes her to get there will depend on how fast she travels. If she drives an average of 50 miles per hour, the trip will take 6 hours. If she stops to visit all the historical sites and enjoy the scenery, she might average only 15 miles an hour; in this case she would need 20 hours on the road for her trip. There are obviously many other possibilities for her average driving speed and time on the road; the important thing to observe is that there is a relationship between the average driving speed and the amount of time she spends on the road. For any average speed we can compute a time on the road; different average speeds give different amounts of driving time.

It is possible to picture the relationship between average speeds and driving times for a 300-mile trip using a graph. To do this we represent average speeds as numbers on a horizontal number line, and number of hours on the road as numbers on a different (vertical) number line. The two lines typically cross at a point corresponding to 0 on both lines. To show that traveling an average of 50 miles per hour (mph) will result in a 6-hour trip, first find the point on the horizontal line corresponding to 50 mph. Then refer to the vertical line to find how far above the horizontal line 6 hours is represented. Mark the point that is this far above the 50-mph position on the horizontal line. To illustrate that an average rate of 15 mph results in a 20-hour trip, repeat the following process.

1. Find 15 mph on the horizontal axis.
2. Find 20 hours on the vertical axis.
3. Mark the point above 15 mph that is opposite the 20-hour mark on the vertical axis.

Study the graph in Figure 5.1 to see how these two points are located.

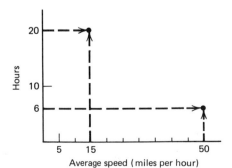

Figure 5.1

Notice that each point corresponds to a pair of numbers. We have found the two points corresponding to (50 mph, 6 hours) and (15 mph, 20 hours). There are many pairs of numbers that give possible average speed and time for a 300-mile trip and, thus, there are many points on our graph. Here are some others.

(10 mph, 30 hours)   (20 mph, 15 hours)

(30 mph, 10 hours)   (60 mph, 5 hours)

We have added the points corresponding to these pairs to our graph (Figure 5.2).

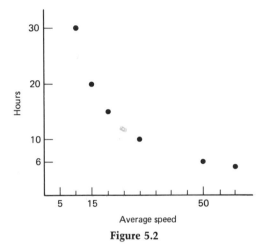

Average speed

Figure 5.2

For each average speed there is a pair of numbers consisting of average speed and time. Of course, we cannot write down all these pairs and find the points on the graph corresponding to them. However, by finding additional points we can see what the graph would look like if we were able to find all the points. Complete these pairs of numbers and add the points corresponding to them to the graph.

Average Speed	Time
12 mph	?
5 mph	?
75 mph	?
25 mph	?

In Figure 5.3, we have sketched a picture of what this graph would look like if we were able to put all the points on the graph that correspond to every possible average speed.

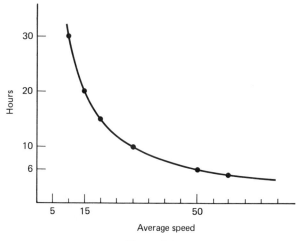

Figure 5.3

Notice that this graph illustrates several features of the 300-mile trip:

1. For any given average speed there is exactly one time required for the trip.
2. The faster the average speed, the fewer hours the trip will take. (This is seen on the graph as follows: the further right on the horizontal axis we locate the average speed, the lower will be the vertical height of the graph.)
3. A small difference between two *slow* average speeds makes a big difference in the amount of time required for the drive.
4. A small difference between two *fast* average speeds makes a *small* difference in the amount of time required for the trip.

The graph also agrees with some obvious truths. For example, it is impossible to make the trip at 0 miles per hour, so there is no point on the graph corresponding to the "speed" 0.

When we graph a relationship between two quantities, we choose scales on the two axes that best accommodate the information that is being pictured. The scales need not be the same and the point of intersection of the two axes need not be labeled 0 on the axes. The graph in Figure 5.4 is from the June 29, 1981, issue of *Newsweek* and shows the number of tourists visiting the national parks in the period 1961 to 1980. The scale on the horizontal axis indicates the various years between 1961 and 1980. Each mark on the vertical scale represents 25,000,000 visits to the national parks. Because the numbers are so large, we are not able to read the graph to find

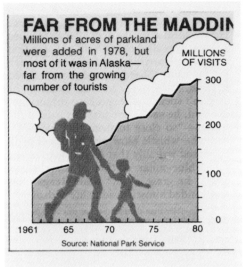

Figure 5.4   NEWSWEEK/JUNE 29, 1981

exactly how many tourists were in the park in a given year. We are able, however, to see that the number has grown steadily over the period (with 1976–77 perhaps being an exception).

PROBLEMS FOR CLASS DISCUSSION

1. Tennis balls are sold in cans containing three balls each. Make a graph showing the relationship between the number of cans a store sells and the number of balls it sells. Use Figure 5.5.

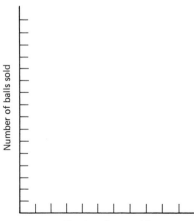

Figure 5.5

2. Suppose that a company can sell its Mighty Wonder Gadget to 100 local families if it charges $100 apiece. For every $10 *increase* in the price of the Wonder Gadget, the company would lose half of its buyers. For every $10 *decrease* in the price, the company would sell twice as many. Make a graph illustrating the relationship between the cost of a Wonder Gadget and the number the company can expect to sell locally.

3. In 1980, the Ohio state income tax was assessed as a percent of total taxable income according to the following tax table. Make a graph illustrating the relationship between taxable income and amount of tax paid.

TAX TABLE

x = income
y = tax paid

Ohio Taxable Income	Formula
$0–5000 Y = .005X	$\frac{1}{2}$% of taxable income
$5000–10,000	$25 plus 1% of excess over $5000
$10,000–15,000	$75 plus 2% of excess over $10,000
$15,000–20,000	$175 plus $2\frac{1}{2}$% of excess over $15,000
$20,000–40,000	$300 plus 3% of excess over $20,000
Over $40,000	$900 plus $3\frac{1}{2}$% of excess over $40,000

Exercises 5-A

1. The graph in Figure 5.6 shows the growth in a savings account when interest is computed at 12 percent compounded monthly and $1000 is the initial deposit.
   (a) Approximately how much money is in the account after 2 years?
   (b) How long must the money be left on deposit in order for the amount to reach $2000?

2. *Newsweek*, August 9, 1982, contained the following graph (Figure 5.7) showing the growth since 1971 in Japanese exports of high-technology products.
   (a) The value of the 1981 exports were how many times greater than the value of the 1971 exports?
   (b) In what year did the value of the exports first exceed $20 billion?

3. The telephone directory gives the charges for a long-distance station-to-station call from Columbus to Chicago as $1.95 for the first three min-

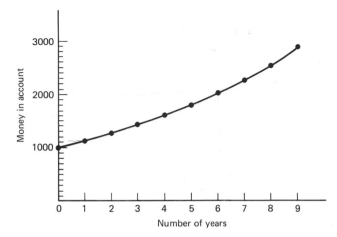

Figure 5.6

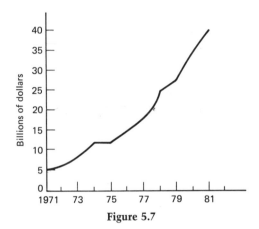

Figure 5.7

utes and $0.32 for each minute or fraction of a minute beyond the first three. Draw a graph that shows the relationship between the length of a call and its cost. Do this for calls up to 10 minutes in length.

4. Garland's Gift Shoppe is having a going-out-of-business sale. Every item is reduced 20 percent. Draw a graph that shows the sale price for every possible original price up to $100.

5. A small town in Florida is growing at an annual rate of 10 percent. If the population in 1980 was 4500 and the annual growth rate continues at 10 percent, draw a graph that shows the population in each year from 1980 to 1990.

6. For all rectangles of a perimeter of 20 feet, draw a graph that shows how the width of the rectangle depends on the length.

7. A culture contains 10 cells. Each cell splits into 2 cells once every half-hour. Draw a graph that shows how the number of cells in the culture depends on the time of the observation.

8. The Athletic Boosters buy pompoms for 65¢ and sell them for $1. They invest in 1000 pompoms. Draw a graph showing the profit for various numbers of pompoms sold.

5.2   GRAPHING EQUATIONS IN TWO VARIABLES

In each of the problems in Section 5.1, there were two variable quantities. We were able to picture the relationship between these quantities by referring to two number lines, each representing the values assumed by one variable, and by sketching a graph. An equation in which two variables appear is a way to describe a relationship between two variables. A graph can provide a geometric picture of all solutions to an equation in two variables.

Here are some examples of equations in two variables.

$$y = x^2 \qquad y = 2x - 1 \qquad x^2 + xy = 4$$

$$s = t^2 + 1 \qquad 5t^3 = 2s^2 - t$$

A solution to an equation in two variables is a pair of numbers. For example, we have a solution to $y = x^2$ when $x = 2$ and $y = 4$, or when $x = -3$ and $y = 9$. In fact, $y = x^2$ has a great many solutions, each of which is a pair of numbers.

We want to draw a picture of all the solutions to the equation $y = x^2$. This picture of the solutions we will call the *graph* of $y = x^2$. In order to represent a solution (a, b) as a point, we take two number lines and position them at right angles as shown. The pair of numbers (a, b) then corresponds to the point that is both opposite the label a on the horizontal number line and opposite the label b on the vertical number line. For example, the pair $(2, 1)$ identifies the point P shown in Figure 5.8 and the pair $(-3, -2)$ identifies the point Q.

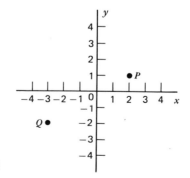

Figure 5.8

We need some vocabulary associated with graphs. Each number line is called an **axis** and the point where they intersect, the **origin.** The pair of numbers that identify the point are called its **coordinates.**

With this introduction we can begin to draw the graph of $y = x^2$. First, we list many solutions giving the x value first and the y value second.

$$(3, 9) \quad (1, 1) \quad (0.5, 0.25) \quad (0, 0) \quad (-0.1, 0.01)$$

$$(-0.7, 0.49) \quad (-1, 1) \quad (-2, 4) \quad (-4, 16)$$

There is no reason for choosing these particular solutions out of all the solutions. We need to list enough so that they give us a good start on the graph of $y = x^2$. The horizontal line, or x axis, is used to locate the x value of a solution and the vertical line, or y axis, is used to locate the y value. When we locate (or plot) a point for each of these solutions, we have a collection of points that looks like Figure 5.9.

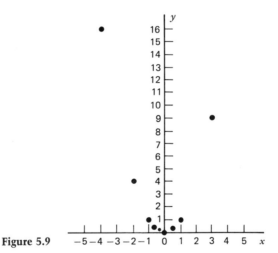

Figure 5.9

If we were able to plot a point for each solution to $y = x^2$, these points would form a curve. It is not clear that we have enough points to know exactly what the shape of the graph will be. You should take a few minutes to find six more solutions to $y = x^2$, plot a point for each of these solutions, and sketch what you think the graph will look like.

Let's consider a second equation in two variables: $y = 2x - 1$. To get a good start on the graph of this equation, we again list several solutions.

$$(3, 5) \quad (2, 3) \quad (1.5, 2) \quad (0.5, 0)$$

$$(0, -1) \quad (-1, -3) \quad (-2, -5)$$

Plotting each of these solutions gives us the picture in Figure 5.10.

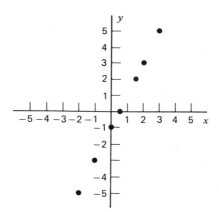

Figure 5.10

This time it is tempting to guess that the solutions to $y = 2x - 1$ all lie on the same line. Plot a few more solutions to convince yourself that this is a reasonable assumption. Is it surprising that the solutions of a *linear* equation give a graph that is a line?

We have already seen examples where a problem situation can be described by a linear equation in two variables. Here is such a situation.

In the office coffee corner, coffee is selling for 20¢ a cup and tea for 5¢ a cup. On Monday, $5.20 was collected. What possibilities are there for the number of cups of coffee sold and the number of cups of tea sold?

There are many possibilities: everything from all coffee to all tea. If x represents the number of cups of coffee and y the number of cups of tea, then $0.20x + 0.05y = 5.20$. We can graph several solutions (Figure 5.11).

$$(0, 104) \quad (10, 64) \quad (15, 44) \quad (25, 4)$$

You can still add additional points to this graph. However, the values for x and y can only be natural numbers or 0 since x and y represent cups of coffee and tea. Thus, the graph will not be a connected curve but rather 27 different points. Yet, these points all lie on the same line as you perhaps have suspected.

We want to consider the graph of one more equation: $y = \dfrac{1}{x}$. In order to investigate points on this graph, we need to use fractions. In

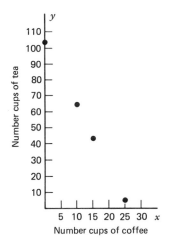

Figure 5.11

Number cups of coffee

the previous chapters we have usually needed only integers in our computations but now we need to make several assumptions about computing with fractions. Appendix D has been written to provide a review of fractions for students who feel rusty. You are encouraged to make a digression to this appendix before going ahead with this section.

We start by listing several solutions to $y = \dfrac{1}{x}$.

$$\left(-5, -\frac{1}{5}\right) \quad \left(-2, -\frac{1}{2}\right) \quad (-1, -1)$$

$$\left(-\frac{1}{2}, -2\right) \quad \left(-\frac{1}{3}, -3\right) \quad \left(-\frac{1}{10}, -10\right)$$

$$\left(\frac{1}{10}, 10\right) \quad \left(\frac{1}{3}, 3\right) \quad \left(\frac{1}{2}, 2\right)$$

$$(1, 1) \quad \left(2, \frac{1}{2}\right) \quad \left(5, \frac{1}{5}\right)$$

Observe that there is no value for y corresponding to $x = 0$ because $\dfrac{1}{0}$ is not a number. Thus, the graph of this equation cannot have a point where $x = 0$. Plotting the points we have identified, gives us the graph in Figure 5.12, on the next page.

Remembering that this graph has no point for $x = 0$, we can sketch in the curve that our points suggest. Do you see that there is a point on this graph for every value of x except $x = 0$ (Figure 5.13)?

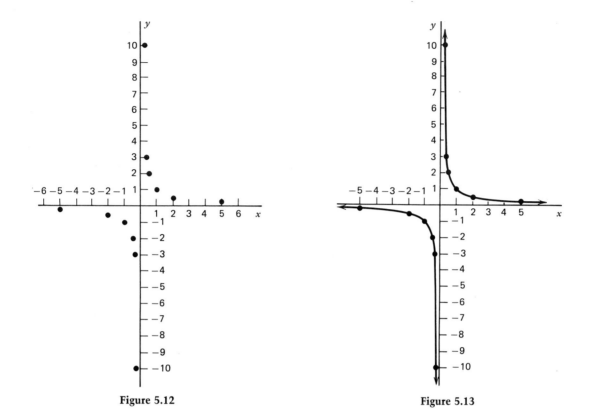

Figure 5.12   Figure 5.13

$$y = \frac{1}{x}$$

Exercises 5-B

In these exercises and in all exercises with graphs, use graph paper and be as accurate as possible.

1. Identify the coordinates of each of the points plotted in Figure 5.14.

2. Label a pair of axes for graphing. Then plot the following points and label each.

$$(0, 0) \quad (5, 0) \quad (-7, 2)$$
$$(-3, 0) \quad \left(\frac{1}{3}, -2\right) \quad \left(3, 3\frac{1}{2}\right)$$

3. Find 10 solutions to $y = x(x + 2)$; then label a pair of axes and carefully plot your solutions.

4. Find 10 solutions to $4y = x^2$; then label a pair of axes and carefully plot your solutions.

5. Find 10 solutions to $2x + y = 3y$; then label a pair of axes and carefully plot your solutions.

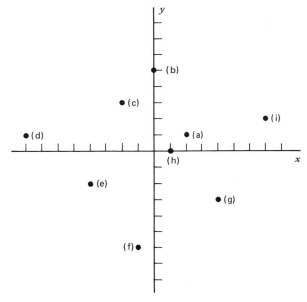

Figure 5.14

6. **(a)** Graph the equation $y = 2x - 1$ using seven points—all different from the points we used on page 74.

   (b) Using the same axes as in part (a), graph the following three equations using at least seven points for each.
   $$y = 2x$$
   $$y = 2x + 2$$
   $$y = 2x - 2$$

   (c) Describe the graphs you have sketched in part (b). *parallel*

7. The sketch in Figure 5.15 shows two graphs. One is the graph of $y = 2x$ as labeled. What is the equation of the other graph?

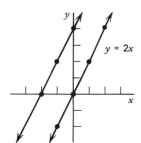

$y = 2x$

Figure 5.15

8. Sketch the graphs of $y = 2x$ and $y = -2x$. How are they different?

9. Sketch the graphs of $y = 2x$ and $y = -\frac{1}{2}x$. What is the relationship between these two graphs?

10. Think about the graph of $y = \frac{1}{x-1}$. For what values of x will there be no corresponding value of y? List eight different solutions for $y = \frac{1}{x-1}$, choosing at least four values for x between 0 and 2. Plot the points representing these solutions and sketch what appears to be the graph.

11. Repeat problem 10 for the equation $y = \frac{1}{x+1}$ choosing at least four values of x between 0 and -2.

12. Graph the equation $2x - y = 1$. Compare it to our graph of $y = 2x - 1$ on page 74. Comment on your observations.

13. Here is the graph of an equation we don't know (Figure 5.16).

    (a) Using this graph, give the y value that the equation assigns to each of the following x values (estimate if necessary).

    $$x = 0 \qquad x = -1 \qquad x = 1 \qquad x = 2$$

    $$x = \frac{1}{2} \qquad x = -\frac{1}{2} \qquad x = 4 \qquad x = -2$$

    (b) Give the x value that corresponds to each of the following y values.

    $$y = -1 \qquad y = 0 \qquad y = 1 \qquad y = 3$$

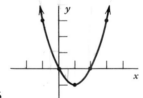

Figure 5.16

14. Compare the graphs of the following four equations. Tell how they are similar and how they are different.

$$y = 3x \qquad y = -x \qquad y = -5x \qquad y = \frac{2}{3}x$$

15. Not every graph is a curve. Whole regions may be graphs for certain inequalities. (See Sections 3.3 and 3.4.) Sketch the graphs of the following inequalities.

    EXAMPLE: All points (x, y) where $x \geq 3$ and $y \leq 4$ are graphed in Figure 5.17.

    (a) All points (x, y) where $x \leq -2$ and $y \leq 0$.
    (b) All points (x, y) where $x > 0$.
    (c) All points (x, y) where $x = 3$.
    (d) All points (x, y) where $y = -2$.

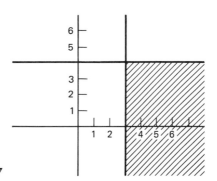

Figure 5.17

16. (a) Graph $y = x^2 - 3x - 4$ by plotting many different points.
    (b) Where does your graph cross the horizontal axis? Are these solutions to $x^2 - 3x - 4 = 0$? Explain.
    (c) Where does your graph have height 6? Are these values solutions to $x^2 - 3x - 4 = 6$? Explain.

17. The graphs of the following equations will require plotting a large number of points. Graph these equations for values of x between -3 and 3.
    (a) $y = x(x + 2)(x - 2)$.
    (b) $y = \dfrac{1}{x - 2}$.

18. In problem 3 of Exercises 1-A, you wrote equations to describe three situations. Sketch graphs of these equations.

19. In Part III of Exercises 1-C, you wrote equations to describe the relationship between two given numbers. Use these pairs of numbers to sketch a graph of each of those equations.

REVIEW PROBLEMS

1. A grocery store is having a "buy two, get one free sale" on all canned goods. Make a graph showing the relationship between the original price of three cans and the cost if you buy them on the sale.

2. Label a pair of axes and plot the following points.
$$(-3, 0) \quad (-2, -7) \quad (0, 6) \quad (0, 0) \quad (4, -4)$$

3. Find 10 solutions to $y = 3x^2$. Label a pair of axes and plot your solutions.

4. Sketch the graph of $y = 4x + 3$.

5. List at least 10 solutions to $y = \dfrac{1}{2x}$ and sketch the graph. For what value(s) of x is there no point on the graph?

$y = 2 \cdot x = \underline{\quad} \ ^1\!/\!x$

CHAPTER 6

LINEAR EQUATIONS IN TWO VARIABLES AND THEIR GRAPHS

6.1   LINEAR EQUATIONS AND LINES

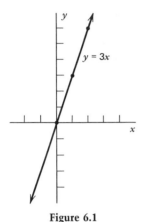

Figure 6.1

The work of the last chapter indicated the graphs of equations are quite varied. In this chapter we focus on equations whose graphs are straight lines—that is, linear equations in two variables.

A graph that is a straight line is easy to draw. In fact, if you know two points that are on the graph, you can draw the whole line; it is not necessary to plot a large number of points. Or, if you know where the graph crosses the vertical axis and the horizontal axis, you can draw the line without having additional information about the graph. Another two characteristics that describe a line are (1) where it crosses the vertical axis and (2) how much it deviates from being horizontal (how much it "slopes"). These two characteristics can be measured numerically and the measurements used in an equation describing the line. These are the two characteristics that we investigate carefully.

Figure 6.2

LINES THROUGH THE ORIGIN

Graphs of equations like $y = 3x$, $y = 10x$, $y = \frac{1}{2}x$ are particularly simple. Each is a line that contains the point $x = 0$, $y = 0$. Thus, each is a line through the origin. Consider first the graph of $y = 3x$. To graph $y = 3x$, we graph all the pairs that can be formed by taking a number and 3 times the number—that is, pairs $(x, 3x)$ for any number x. Three such points are $(0, 0)$, $(1, 3)$, $(2, 6)$. Take a look at Figure 6.1. Notice that the effect of the multiplier 3 in the equation $y = 3x$ is to lift the line as we go to the right.

Consider next the graph of $y = 1x$ (Figure 6.2). The multiplier 1 in this equation is positive and is less than the multiplier 3 in the previous example. Its graph consists of points (x, x) for any number x. Three such points are $(0, 0)$, $(1, 1)$, $(2, 2)$. The multiplier 1 lifts the line as we move from left to right but does not lift it as rapidly as the multiplier 3 in the previous example. The line with the larger multiplier is steeper.

Now you can guess that the graph of $y = 2x$ will lie between the graphs of $y = 1x$, and $y = 3x$. When the graphs of $y = 1x$, $y = 2x$, $y = 3x$, and $y = 5x$ are drawn on the same axes, the effect of the multiplier of x on the graph of the equation becomes more clear (Figure 6.3).

We need also to consider lines through the origin that are described by equations with negative multipliers of x. Consider $y = -3x$. Three points on this line are $(0, 0)$, $(1, -3)$, and $(2, -6)$. Notice

Figure 6.3

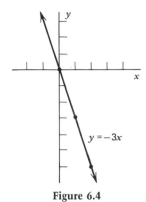

Figure 6.4

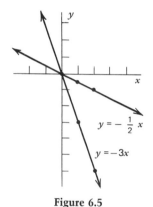

Figure 6.5

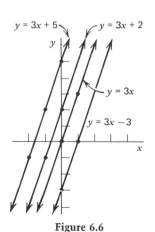

Figure 6.6

that the effect of the negative multiplier is to cause the line to fall rather than rise as we go right (Figure 6.4).

Any negative multiplier will cause the line to fall as we move to the right. Consider $y = -\frac{1}{2}x$. By graphing the points $(0, 0)$, $\left(1, -\frac{1}{2}\right)$, $(2, -1)$, we see that the line goes down as we move from left to right but not as much as in the graph of $y = -3x$ when the two lines are drawn on the same coordinate system (Figure 6.5).

Lines through the origin (except for the y axis) are described by equations of the form $y = mx$. If the multiplier m is a positive number, the line rises when we go from left to right; if the multiplier m is negative, the line falls. If the multiplier is 0, the equation is $y = 0$ and it describes the horizontal axis. Remember, if (x, y) is any point on the horizontal axis, then $y = 0$.

LINES NOT NECESSARILY THROUGH THE ORIGIN

Of course, not every line in the planes goes through the origin. However, if you have a line that does not go through the origin, you can draw a second line parallel to it that does go through the origin. This is why we study families of parallel lines. For example, in addition to the equation $y = 3x$ we can also consider the equations $y = 3x + 2$ and $y = 3x + 5$ and $y = 3x - 3$. You graphed families of lines like this in the last chapter. The line described by $y = 3x + 2$ can be viewed as the line $y = 3x$ moved up 2 units. The line $y = 3x + 5$ is the line $y = 3x$ moved up 5 units. And the line $y = 3x - 3$ is the line $y = 3x$ moved down 3 units. These lines are parallel because their equations all have the same multiplier of x. Observe also that the line $y = 3x$ crosses the vertical axis at 0; $y = 3x + 2$ crosses the vertical axis at 2; $y = 3x + 5$ crosses at 5; $y = 3x - 3$ crosses at -3 (Figure 6.6). Each of these equations contains two important numbers that describe the graph of the equation. We have names for these numbers.

The number m in the equation $y = mx + b$ is called the **slope** of the line described by the equation. The number b is called the **y intercept** of the line. The slope tells us how steep the line is and whether it slants up or down. The y intercept tells us where the line crosses the vertical axis, that is, the value of y when $x = 0$.

But there are linear equations in two variables that do not have the form $y = mx + b$. An example is $3x + 4y = 9$. Is it possible that its graph is the same as those we have been examining? To answer this equation, we change the form of the equation and see that it is equivalent to an equation of the form $y = mx + b$:

$$3x + 4y = 9$$
$$4y = -3x + 9 \qquad \text{(Subtract } 3x \text{ from both sides)}$$
$$y = -\frac{3}{4}x + \frac{9}{4} \qquad \text{(Divide both sides by 4)}$$

Thus, the equation $3x + 4y = 9$ is equivalent to $y = -\frac{3}{4}x + \frac{9}{4}$. Its graph is a line with slope $-\frac{3}{4}$ and y intercept $\frac{9}{4}$.

In this example, rewriting the equation in the form $y = mx + b$ required that we divide by 4 in the last step. If the coefficient of y in the original equation had been 0 rather than 4, we could not have solved for y. Indeed, there would have been no y appearing in the equation! Thus, we want to look at two special cases of linear equations. The first has the coefficient of y equal to 0; the second has the coefficient of x equal to 0.

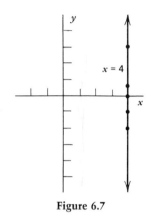

Figure 6.7

TWO SPECIAL CASES

First, consider equations that look like $x = a$ for some number a. We can regard the coefficient of y as 0. An example is the equation $x = 4$. To graph this equation we graph all points (x, y) that have $x = 4$. Here are some solutions: $(4, -2)$, $(4, -1)$, $(4, 0)$, $\left(4, \frac{1}{2}\right)$, $(4, 3)$. The collection of all points satisfying $x = 4$ is the line parallel to the vertical axis that goes through 4 (Figure 6.7). Similarly the graph of $x = -1$ is the line parallel to the vertical axis that goes through -1. The vertical axis itself has equation $x = 0$. Equations of the form $x = a$ cannot be written in the form $y = mx + b$. We say that the slope of these lines parallel to the vertical axis is **undefined.**

As a second special case, consider equations of the form $y = b$ for some number b. In this equation we can regard the coefficient of x as 0 because $y = b$ is equivalent to $y = 0x + b$. An example is $y = -2$. A list of solutions for $y = -2$ includes $(-1, -2)$, $\left(-\frac{3}{4}, -2\right)$, $(0, -2)$, $(1, -2)$, $(5, -2)$. The collection of all solutions to $y = -2$ is the line parallel to the horizontal axis that goes through -2 (Figure 6.8). Similarly the graph of $y = \frac{5}{2}$ is the line parallel to the horizontal axis that goes through $\frac{5}{2}$. The horizontal axis has equation $y = 0$. Since equations of the form $y = b$ can be written as $y = 0x + b$, they have slope 0. Thus, all lines parallel to the horizontal axis have slope 0.

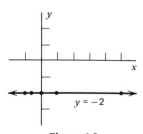

Figure 6.8

Exercises 6-A

I. For each of the following equations, give the slope and the y intercept.

1. $y = 3x + 1$

2. $y = 3x - 1$

3. $3x + y = 4$

4. $-5x + 2y = 20$

5. $8x + 2y = 0$

6. $y = -5$

7. $x = -5$

8. $2y = 4$

9. $x + 4y - 8 = 0$

10. $3x - 5y + 15 = 0$

II. We have said the y intercept of a line is the point where the line crosses the vertical axis. The **x intercept** is the point where the line crosses the x axis. Since the x intercept has y coordinate 0, it can be found by setting $y = 0$ and solving for x. Given the x intercept and the y intercept for each of the following and sketch the line.

EXAMPLE: $3x + y = 9$
For x intercept: $3x + 0 = 9$
$3x = 9$
$x = 3$
For y intercept: $3 \cdot 0 + y = 9$
$y = 9$

1. $2x + 3y = 6$

2. $x - y = 2$

3. $y = \frac{1}{2}x + 3$

4. $-2x = y + 1$

III. Write an equation for the line having each of the following properties.

1. Slope $= 2$; y intercept $= 1$.

2. Slope $= -3$; y intercept $= 5$.

3. Slope $= \frac{1}{2}$; y intercept $= 0$.

4. Slope $= 100$; y intercept $= 25$.

5. Slope $= -9$; y intercept $= -3$.

6. Slope $= \frac{3}{4}$; y intercept $= -\frac{1}{2}$.

7. Slope $= 0$; y intercept $= 4$.

8. Slope $= 0$; y intercept $= -2$.

9. Slope undefined; crosses x axis at $\frac{3}{2}$.

10. Slope undefined; crosses x axis at -3.

11. The horizontal axis.

12. The vertical axis.

6.2   GEOMETRIC INTERPRETATION OF SLOPE

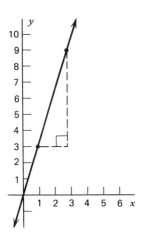

Figure 6.9

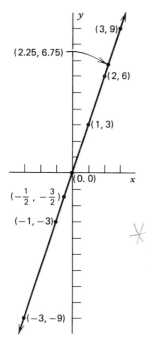

Figure 6.10

We have considered the special linear equations $x = a$ and $y = b$ for numbers a and b. Now we return to equations of the form $y = mx + b$. The number b is called the y intercept of the graph of the equation and it indicates the point on the vertical axis where the line crosses. The number m is called the slope of the line and it somehow describes how the line slants. We can get a more precise geometric meaning for slope by examining the examples closely.

Consider again the line $y = 3x$ and two points $(1, 3)$ and $(3, 9)$ on the line. Using the two points for vertices as shown in Figure 6.9, we form a right triangle whose height is $9 - 3 = 6$ and whose base has length $3 - 1 = 2$. The height of the triangle is three times the length of the base. The same is true, in fact, if we repeat the procedure for any two points on the graph of the line $y = 3x$. Several of these points are given on the graph in Figure 6.10. If instead of $(1, 3)$ and $(3, 9)$ we choose $(-1, -3)$ and $(2.25, 6.75)$, the base of the triangle has length $2.25 - (-1) = 3.25$ and the height of the triangle has length $6.75 - (-3) = 9.75$ (Figure 6.10). Notice that $9.75 = 3 \cdot (3.25)$, and again the height of the triangle is three times the base. You should try this yourself for several other pairs of points given on the graph and convince yourself that the height will always be 3 times the length of the base.

The example in Figure 6.10 is a line with positive slope 3. Consider an example with negative slope: $y = -2x$. If we take two points, say $(1, -2)$ and $(2, -4)$, on the line $y = -2x$, and form a right triangle as before (Figure 6.11), then the difference of the second coordinates can be written as $-4 - (-2)$ or -2 and the difference of the first coordinates as $2 - 1 = 1$. The quotient is the slope: $\dfrac{-2}{1} = -2$. In this case the slope indicates that the line drops 2 units when we move one unit to the right. Rather than taking the particular points $(1, -2)$ and $(2, -4)$, we might have taken any two points on the line. If we compute the difference of the second coordinates divided by the difference of the first coordinates, the result will be the same—the slope -2. You should identify other points on the line and try this. Notice that we remember which point comes first when we compute the differences of the coordinates, so that the difference of the second coordinates and the difference of the first coordinates are taken in the same order.

In short, if (a, b) and (c, d) are two points on a line not parallel to the vertical axis, the line will have slope

$$m = \frac{b - d}{a - c} \quad \text{or} \quad m = \frac{d - b}{c - a}$$

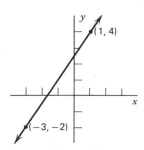

$y = -2x$

$(1, -2)$

$(2, -4)$

Figure 6.11

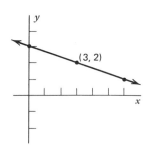

$(1, 4)$

$(-3, -2)$

Figure 6.12

The quotient gives $\dfrac{\text{change in } y}{\text{change in } x}$. For example, the points $(1, 4)$ and $(-3, -2)$ lie on a line that has a slope $\dfrac{4 - (-2)}{1 - (-3)} = \dfrac{4 + 2}{1 + 3} = \dfrac{6}{4} = \dfrac{3}{2}$. If we had taken the points in the other order, we would have computed $\dfrac{-2 - 4}{(-3) - 1} = \dfrac{-6}{-4} = \dfrac{3}{2}$ (Figure 6.12).

It is possible to graph a line if we know one point on the line and the slope. This can be done without knowing the equation of the line. For example, say you want to sketch the graph of the line through the point $(3, 2)$ with slope $-\dfrac{1}{3}$. First, locate $(3, 2)$. Then, move down 1 and right 3 to locate a second point. (Remember, the numerator of the slope describes vertical change; the denominator describes horizontal change.) If you had regarded the slope as $\dfrac{1}{-3}$, you would have moved 1 unit up and 3 units left to locate another point on the line. Either way, you have located two points on the line you want to graph.

Knowing one point on a line and the slope is also enough information to write the equation of the line. In the example in Figure 6.13, we have slope $-\dfrac{1}{3}$ and point $(3, 2)$. Since the line is not parallel to an axis, its equation must be $y = mx + b$ where the slope $m = -\dfrac{1}{3}$. Thus, we know the equation looks like $y = -\dfrac{1}{3}x + b$; finding a numerical value for the y intercept b is all that remains to be done. Recall that since $(3, 2)$ is a point on the line, the pair of values $x = 3$, $y = 2$ is a solution to the equation $y = -\dfrac{1}{3}x + b$. This means that

$$2 = \left(-\frac{1}{3}\right)(3) + b$$

Solving this equation for b gives $b = 3$. Therefore, the equation of the line through the point $(3, 2)$ with slope $-\dfrac{1}{3}$ is $y = -\dfrac{1}{3}x + 3$.

The preceding example provides you with a method of finding an equation for a line if you know two points on the line. First use the two points to determine the slope of the line (denoted by m). Write the equation in the form $y = mx + b$. Then use either point to assign values to x and y; this will give a linear equation in the single variable b. Solve it! Knowing the values of m and b, you can write the equation of the line.

$(3, 2)$

Figure 6.13

SUMMARY

In this section we have said many things about linear equations and lines. Here are some of the important ideas.

The graph of an equation of the form $y = mx + b$ is a line with slope m and y intercept b.

The graph of an equation of the form $x = a$ is a straight line parallel to the vertical axis; it has undefined slope.

The graph of an equation of the form $y = b$ is a straight line parallel to the horizontal axis; it has slope 0.

If (a, b) and (c, d) are two points on a line, the slope of the line is $m = \dfrac{b - d}{a - c}$.

If a line passes through a point (u, v) and has slope m, its equation is $y = mx + b$ where the number b can be found by replacing y in the equation by the coordinate v, and x by u.

An equation of the form $cx + dy = n$ where $c \neq 0$ and $d \neq 0$ can always be written in the form $y = mx + b$, and the slope and y intercept can be determined from the second form.

WRITING EQUATIONS OF LINES

We complete this section by considering some typical examples in which equations are written to describe particular lines. Study these examples and in each case sketch the graph of the line.

EXAMPLE 1   Find the equation of the line containing the points $(-1, 2)$ and $(1, 3)$.

The slope of this line is $\dfrac{3 - 2}{1 - (-1)} = \dfrac{1}{2}$. Thus, its equation has the form $y = \dfrac{1}{2}x + b$. Since $(-1, 2)$ is a solution of this equation, we can write

$$2 = \frac{1}{2}(-1) + b$$

$$2 = -\frac{1}{2} + b$$

$$2\frac{1}{2} = b$$

The equation, therefore, must be $y = \dfrac{1}{2}x + 2\dfrac{1}{2}$. Notice that we could also have used the point $(1, 3)$ to solve for b in the equation

$$y = \frac{1}{2}x + b$$

$$3 = \frac{1}{2}(1) + b$$

$$2\frac{1}{2} = b$$

EXAMPLE 2 Find the equation of the line containing the point $(2, -5)$ with slope 0.

This equation must have the form $y = 0x + b$. Since $(2, -5)$ is a solution, we can replace x by 2 and y by -5 to get $-5 = 0(2) + b$, or $-5 = b$. The equation becomes $y = 0x - 5$ or $y = -5$. This is a line parallel to the horizontal axis.

(handwritten: $-5 = 0(2) + b$ $-5 = 0 + b$ $-5 = b$*)*

EXAMPLE 3 Find the equation of the line through point $(-2, 3)$ and parallel to $x + y = 4$.

(handwritten: $y = -x + 4$ $3 = -1x + 4$*)*

Recall that parallel lines have the same slope; therefore, the line we want an equation for has the same slope as the line described by $x + y = 4$. To find that slope, we rewrite the equation as $y = -x + 4$. Now we know the line has slope -1. Thus, we are looking for a line with slope -1 containing the point $(-2, 3)$. The form of the equation is $y = -x + b$. We find b by evaluating our equation at $(-2, 3)$.

(handwritten: $y = mx + b$ $3 = (-1)(-2) + b$ $3 = 2 + b$ $1 = b$*)*

$$y = -x + b$$
$$3 = -(-2) + b$$
$$3 = 2 + b$$
$$1 = b$$

(handwritten: because both are parallel we can use the same equation*)*

Thus, the equation is $y = -x + 1$.

Exercises 6-B

I. Graph each of the following without writing an equation for the line.
1. The line through $(1, -1)$ and $(4, 3)$.

2. The line through $(2, 3)$ with slope $\frac{1}{2}$.

(handwritten: book graph is wrong*)*

3. The line through $(-1, 2)$ with slope $-\frac{1}{2}$.

4. The line through $(4, -5)$ with slope 5.

5. The line through $(-1, -1)$ with slope -3.

6. The line through $(2, -3)$ parallel to the horizontal axis.

7. The line through $(1, 4)$ parallel to the vertical axis.

II. Graph each of the equations in Part III of Exercises 6-A.

III. Find the slope of the line determined by each of the following pairs of points.

1. $(4, 3)$ and $(2, 1)$.

2. $(5, -1)$ and $(3, -4)$.

3. $(-3, -2)$ and $(11, -10)$.

4. $(-10, 11)$ and $(11, -10)$.

5. $(6, -5)$ and $(4, -2)$.

6. $(-7, 1)$ and $(-8, 5)$.

7. $(3, 4)$ and $(-3, 4)$.

8. $(5, 7)$ and $(5, -7)$.

9. $(4, -8)$ and $(-4, 8)$.

10. $(7, -3)$ and $(-6, -4)$.

11. $(0, -3)$ and $(-6, 0)$.

12. $(0, 0)$ and $(4, -7)$.

IV. Solve the following problems.

1. In our discussion we have said that equations like $x = 2$ have *undefined slope*. Choose two points on the graph of $x = 2$ and attempt to compute the slope. What happens?

2. In our discussion we have said that equations like $y = 2$ have slope 0 (since $y = 2$ is the same as $y = 0 \cdot x + 2$). Choose two points on the graph of $y = 2$ and compute the slope. Do you get 0?

V. Find an equation of the line with the given slope through the given point.

1. Slope $\frac{1}{2}$ through $(0, 5)$.

2. Slope -3 through $(11, 7)$.

3. Slope 0 through $(4, 3)$.

4. Slope 4 through $(2, 2)$.

5. Slope $-\frac{2}{3}$ through $(-2, -3)$.

6. Slope $\frac{2}{3}$ through $(-4, 5)$.

7. Slope 10 through $(4, 0)$.

8. Slope $\frac{7}{2}$ through $(1, -2)$.

9. Slope $\frac{4}{3}$ through $(-2, 4)$.

10. Slope $-\frac{2}{5}$ through $(2, -5)$.

VI. Find an equation for a line that has each of the following properties. Graph the line.

1. A line through $(4, 2)$ and $(1, 1)$.

2. A line through $(3, 5)$ and $(-1, 0)$.

3. A line through $(-1, -7)$ and $(-3, -4)$.

4. A line through $(3, 3)$ and $(3, 8)$.

5. A line through $(4, -2)$ and $(0, 2)$.

6. A line that crosses the horizontal axis at 5 and the vertical axis at -2.

7. A line parallel to the horizontal axis containing the point $(-1, 1)$.

8. A line parallel to the vertical axis containing the point $(-1, 1)$.

9. A line parallel to $2x + 5y = 80$ and containing $(-1, 2)$.

10. A line with slope $-\frac{1}{2}$ and containing the origin. $y = -\frac{2}{5}(x) + \frac{8}{5}$

VII. Solve the following problems.

1. To convert temperature from Celsius measure to Fahrenheit, you must multiply the number of degrees Celsius by $\frac{9}{5}$ and add 32. Write an equation that describes this conversion, graph the equation, and give the slope and y intercept.

2. On Up-Up-And-Away Day, the county airport offers half-hour airplane rides at a cost of a penny per pound plus $2. Write an equation that shows how cost is determined from a passenger's weight, graph the equation, and give the slope and y intercept.

3. The Little Licorice Company figures that its daily costs are $0.03 for each foot of licorice made plus $120. Write an equation that shows how to compute the daily costs in terms of the number of feet of licorice produced, graph the equation, and give the slope and y intercept.

6.3 SOLVING TWO SIMULTANEOUS LINEAR EQUATIONS

In the previous chapter we discussed several problem situations that could be described by one linear equation. Some problem situations are naturally described by more than one linear equation. Consider this problem.

The refreshment committee for a community function sold steins of beer for $1 each and coffee for 50¢ per cup. During the evening they sold twice as many steins of beer as cups of coffee and collected $200. How many of each did they sell?

To solve this problem we might let x represent the number of cups of coffee sold and y the number of steins of beer. Then the problem tells us two things.

1. $y = 2x$ (Twice as many beers as coffees)
2. $0.50x + 1.00y = 200$ (Total amount collected)

We must find a pair of numbers that is a solution for both of these equations. In cases like this, we are solving **simultaneous equations;** we look for values of the variables x and y that serve simultaneously as solutions to both equations.

Since the first equation tells us $y = 2x$, we can replace y by $2x$ in the second equation, giving

$$0.50x + 1.00(2x) = 200$$

$$0.50x + 2.00x = 200$$

$$2.50x = 200$$

$$x = 80$$

The computation indicates that the committee sold 80 cups of coffee. Since $y = 2x$, they must also have sold 160 steins of beer. You should test these solutions to make sure they solve both equations.

The simultaneous equations we just solved were of a special form: the first equation gave one variable in terms of the other and it was natural to replace that variable in the second equation. This procedure is sometimes called **substitution.** Here is another case of two simultaneous equations that can be solved readily by substitution.

$$2x + 3y = 13$$

$$x + 2y = 8$$

The second equation is equivalent to $x = 8 - 2y$ (subtract $2y$ from both sides). Thus, we can replace x in the first equation with $8 - 2y$, giving $2(8 - 2y) + 3y = 13$. This equation has only one variable and can be solved as follows.

$$
\begin{aligned}
2(8 - 2y) + 3y &= 13 \\
16 - 4y + 3y &= 13 \qquad \text{(Distributive property)}\\
16 - y &= 13 \\
-y &= -3 \qquad \text{(Subtract 16 from each side)}\\
y &= 3
\end{aligned}
$$

Remember, we need a value for x as well as for y. Since the second equation says that $x + 2y = 8$, if $y = 3$ then $x + 2(3) = 8$ or $x = 2$. Thus, $(2, 3)$ is a solution to both of the original equations. Check it!

The preceding examples were readily solved by the method of substitution because one variable could easily be expressed in terms of the other. Here is an example where it is not as easy to write one variable in terms of the other.

$$2x - 3y = 5$$

$$4x + 3y = 7$$

Since we seek a pair of numbers (x, y) such that $2x - 3y = 5$ *and* $4x + 3y = 7$, it follows that the sum of the left sides of these equations equals the sum of the right sides; that is,

$$(2x - 3y) + (4x + 3y) = 5 + 7$$

The effect of the addition is to eliminate the y term and give a value for x. Therefore,

$$
\begin{aligned}
2x - 3y + 4x + 3y &= 12 \\
6x &= 12 \\
x &= 2
\end{aligned}
$$

To find the value for y we replace x in one of the original equations by 2.

$$2(2) - 3y = 5$$

$$-3y = 1$$

$$y = -\frac{1}{3}$$

We have $\left(2, -\frac{1}{3}\right)$, a solution for both equations. (You check!)

Again, the above example appears to be special: the first equation contained $-3y$ and the second $3y$ so that, when we added, the y term was eliminated. However, this method (called **eliminating a variable**) can be used more generally. Consider this example.

$$3x + 2y = 19$$

$$2x + 5y = 9$$

Substitution does not seem efficient. Can we eliminate a variable? If we multiply both sides of the first equation by 2 and both sides of the second equation by -3, we get a pair of equations equivalent to the original equations.

$$6x + 4y = 38$$

$$-6x - 15y = -27$$

Now we see that by adding we can eliminate the x term and get $-11y = 11$, so $y = -1$. To find x we replace y by -1 in one of the original equations, say the first.

$$3x + 2(-1) = 19$$
$$3x - 2 = 19$$
$$3x = 21 \qquad \text{(Add 2 to both sides)}$$
$$x = 7$$

Check to see that $(7, -1)$ is a solution to both equations. What the example illustrates is that the method of eliminating a variable can often be used on equations equivalent to the given equations: we multiply the original equations by numbers that will cause one variable to have a coefficient of 0 when the new equations are added together.

But is there always a common solution to any two linear equations? Think about the geometry of the situation. Since the solutions to a linear equation can be pictured as a straight line, we are asking if any two lines in the same plane always have a point in common. This will be true if the lines intersect; then they will have a point in common. The graphs of $3x + 2y = 19$ and $2x + 5y = 9$ from the previous example are shown in Figure 6.14. You can see that they have the common point $(7, -1)$. Indeed, if we were able to graph accurately, graphing would be a good technique for solving two simultaneous equations. However, it is unlikely that we could identify a solution like $\left(3\frac{7}{8}, -\frac{1}{5}\right)$ on graphs, and so we depend on the algebraic methods for precise solutions and use the graphs to help us know what to expect.

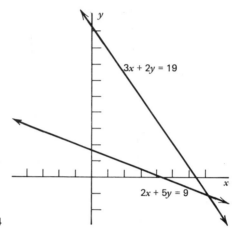

Figure 6.14

Of course, two lines may not intersect. Two linear equations may describe parallel lines or even describe the same line. We can check the slopes of the lines to see if the lines are parallel. If two lines have the same slope but different y intercepts, they will be parallel and have no point in common. For example, consider

$$x + y = 3$$
$$2x + 2y = 8$$

Rewriting these two equations to read the slopes and the y intercepts, we have

$$y = -x + 3$$
$$y = -x + 4$$

Both have slope -1. The first has y intercept 3 and the second has y intercept 4. Thus, they are different lines and are parallel. We can conclude that these two linear equations have no common solutions. If we had failed to notice that the two lines were parallel, we might have attempted to solve the system of equations by multiplying the second equation by -1 and then adding the equations

$$\left.\begin{array}{l} y = -x + 3 \\ y = -x + 4 \end{array}\right\} \rightarrow \left.\begin{array}{l} y = -x + 3 \\ -y = x - 4 \end{array}\right\} \rightarrow \quad 0 = 0 - 1$$

Now we have the equation $0 = -1$, which is never true, and we can conclude "no solutions" from this fact.

Finally, it is possible that two equations that look different actually are equivalent and describe the same line. Consider

$$5x + \quad y = 4$$

$$x + \frac{1}{5}y = \frac{4}{5}$$

Here if both sides of the second equation are multiplied by 5, the result is the first equation. Thus, the two equations are equivalent and every solution to one is a solution to the other. In fact, there is an infinite number of common solutions. Again, if we failed to observe that the two equations have the same graph, we might have first multiplied both sides of the second equation by -5 and then added to eliminate a variable.

$$\left.\begin{array}{l} 5x + \quad y = 4 \\ x + \frac{1}{5}y = \frac{4}{5} \end{array}\right\} \rightarrow \left.\begin{array}{l} 5x + y = \quad 4 \\ -5x - y = -4 \end{array}\right\} \rightarrow 0 = 0$$

The resulting equation is always true, and we can conclude with certainty that each number is a solution.

Here are some additional problem situations that can be described using two linear equations. You should write the equations and find their common solution using the methods discussed above.

PROBLEMS FOR EXAMINATION AND DISCUSSION

1. A candy store is selling two mixtures of valentine candy. One mixture is made by mixing 5 pounds of candy hearts with 5 pounds of cinnamon hearts; it sells for $1.64 per pound. The other mixture is made by mixing 6 pounds of candy hearts and 2 pounds of cinnamon hearts. It sells for $1.47 per pound. What is the price of a pound of candy hearts? Cinnamon hearts? (In this problem we are making the assumption that mixing and measuring candies does not increase the cost of the mixture. Often our textbook examples simplify the real-world situation to make the problem more manageable.

2. Find the number of quarters and dimes in a collection containing 22 coins worth $3.55.

3. How many gallons of a solution containing 30 percent alcohol and how many gallons of a solution containing 80 percent alcohol must be mixed to obtain 30 gallons of a 45 percent solution?

Exercises 6-C

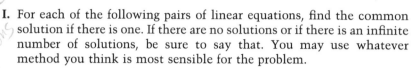

I. For each of the following pairs of linear equations, find the common solution if there is one. If there are no solutions or if there is an infinite number of solutions, be sure to say that. You may use whatever method you think is most sensible for the problem.

1. $x + y = 7$
   $x - y = 3$

2. $2x + y = 5$
   $y = 3$

3. $x + y = 4$
   $x = 5 - y$

4. $3x - y = 5$
   $2y = 6x - 10$

5. $3x + 2y = 19$
   $2x + 3y = 16$

6. $2x + 3y = 13$
   $x + 2y = 8$

7. $2x - 5y = 4$
   $3x + 2y = 25$

8. $2x - 3y = 10$
   $3x - 4y = 11$

9. $5x + 3y = 0$
   $8x + 2y = 0$

10. $3x + 7y = -8$
    $-4x + 5y = 25$

11. $4x + y = 1$
    $6x + 2y = -6$

12. $-5x + 2y = 22$
    $x - y = 7$

13. $3x + 2y + 2 = 0$
    $x - y + \dfrac{3}{2} = 0$

14. $9m + 4n = 6$
    $3m = 5 - 2n$

15. $y = x + 1$
    $y = -\dfrac{1}{2}x + 4$

II. Examine and answer the following questions.

1. In an example on page 94, we analyzed the two equations

$$x + y = 3$$
$$2x + 2y = 8$$

We concluded that their graphs are two different parallel lines. Thus, the graphs have no point in common and the equations have no common solution. Attempt to solve the simultaneous equations and indicate at what point in the computation you are able to conclude that the equations cannot have a simultaneous solution.

2. Also on page 95, we analyzed the equations

$$5x + y = 4$$
$$x + \dfrac{1}{5}y = \dfrac{4}{5}$$

and concluded that the two equations have the same graph. Thus, the equations should have an infinite number of solutions. Attempt to solve these equations and show from the computation that the equations have an infinite number of simultaneous solutions.

III. Solve the following problems.

1. A hardware store sells nuts and bolts by the pound. One customer bought 3 pounds of nuts and 5 pounds of bolts for $21. Another customer bought 5 pounds of nuts and 3 pounds of bolts for $19. What is the price of a pound of nuts? Of a pound of bolts?

2. Football tickets were sold at $4 for an adult ticket and $3 for a child's ticket. Two hundred tickets were sold for a total of $700. How many were adult tickets and how many children's?

3. A packet of 25 stamps costing $3.54 contains both 8¢ and 15¢ stamps. How many of each does it contain?

4. A jet plane traveling with the wind flies 2325 miles in 3 hours. Against the wind it takes the jet 4 hours to go 2900 miles. Find the speed of the plane in still air and the speed of the wind. (It is a fact of physics that if s is the speed of the plane in still air and w the speed of the wind, then $s + w$ is the speed of the plane with the wind and $s - w$ the speed of the plane against the wind.)

5. A plane flies 1200 miles with the wind in 3 hours and then turns around and flies back against the wind in 4 hours. What is the speed of the plane in still air and the rate of the wind?

6. A 40 percent alcohol solution and a 70 percent alcohol solution are to be combined to give 20 gallons of a 52 percent solution. How many gallons of each should be used?

7. A martini is made from gin and vermouth. An 80 proof gin is 40 percent alcohol and vermouth is 18 percent alcohol. How many ounces of each would you mix to get a quart of martinis that is $\frac{1}{3}$ alcohol? (1 quart = 32 ounces.)

8. A rich and creamy salad dressing contains 400 calories per cup. A local dressing contains 50 calories per cup. How would you mix these two dressings to obtain a quart of dressing that has 150 calories per cup? (4 cups = 1 quart.)

IV. Use the following question to generate a class discussion.

Several of the problems in Part III make some assumptions that serve to simplify the situation. Point out what some of these assumptions are.

6.4   THREE OR MORE SIMULTANEOUS LINEAR EQUATIONS

Just as there are problem situations that can be described with two simultaneous linear equations, there are situations that can be described with three or more equations. Here is one that can be stated briefly.

A boy had 30 coins, all nickels, dimes, and quarters. The amount of money in nickels was the same as the amount of money in dimes. The total value of the coins was $3.65. How many coins of each type did he have?

Since there are three numbers to be found, we need three variables. Let

n represent the number of nickels

d represent the number of dimes

q represent the number of quarters

Then the information in the problem can be translated into these equations.

$n + d + q = 30$ (The total number of coins is 30)

$0.05n = 0.10d$ (The value in nickels equals the value in dimes)

$0.05n + 0.10d + 0.25q = 3.65$ (The total value is $3.65)

There are several ways of finding a common solution for these equations. The second equation is equivalent to $n = 2d$ (divide both sides by 0.05). If we replace n in the first and third equations by $2d$, we will have two equations in two variables.

$$2d + d + q = 30$$

$$0.05(2d) + 0.10d + 0.25q = 3.65$$

or

$$3d + q = 30$$

$$0.20d + 0.25q = 3.65$$

This problem can be solved with the methods of the last section for two linear equations in two variables. From the first equation $q = 30 - 3d$, so we substitute for q in the second equation.

$$0.20d + 0.25(30 - 3d) = 3.65$$

$$0.20d + 7.50 - 0.75d = 3.65$$

$$-0.55d = -3.85$$

$$d = 7$$

The number of dimes is 7. Now we look for an equation that will give the number of quarters in terms of the number of dimes. $3d + q = 30$ will do. Replacing d with 7 gives $3(7) + q = 30$ or $q = 9$. We still need the number of nickels but, since we have numerical values for d and q, we can find n using $n + d + q = 30$. Making the substitutions $d = 7$, $q = 9$ gives

$$n + 7 + 9 = 30$$
$$n + 16 = 30$$
$$n = 14$$

Check to see that 14 nickels, 7 dimes, and 9 quarters is a solution to the original problem.

This example illustrates that procedures for solving three linear equations in three variables use the same techniques as two equations in two variables except that the techniques have to be applied more times. In the example we combined equations 1 and 2 by substitution to get an equation in two variables rather than three; then we combined equations 2 and 3 to get an equation in the same two variables. Finally, we solved the two equations in two variables as in the last section. We could have eliminated a variable rather than using the method of substitution. The goal is to combine the three equations in three variables to get two equations in two variables. There are many "right" ways to proceed in finding a solution to three equations in three variables and often no "best" way. Experience is what gives us guidance in how to get started. Remember that a solution will always consist of three numbers.

Here is a second example of how to proceed.

$$2x + y + z = 0$$
$$x + y + 5z = 2$$
$$3x + y - z = 0$$

There are several things that could be done first. We might rewrite the second equation to get $x = -y - 5z + 2$ and then make a substitution for x in both the first and third equations. That would give two equations in y and z, which we could solve by the techniques of the last section. Try that approach.

Another thing we could do in this example would be to multiply both sides of the first equation by (-1) and then add it to the second and third equations. This will remove the y term from both of the sums, leaving two equations in x and z. We pursue that option here.

$$-2x - y - z = 0$$
$$x + y + 5z = 2$$
$$3x + y - z = 0$$

These are the equations after the first of the original equations is multiplied by -1. By adding equations 1 and 2 and then by adding equations 1 and 3, we obtain

$$-x + 4z = 2$$

$$x - 2z = 0$$

Again, there is more than one thing we could do next, but adding the equations to eliminate the x variable gives us

$$2z = 2$$

Thus, $z = 1$. To get a value for x, we can use the equation $x - 2z = 0$. By letting $z = 1$, we can obtain $x - 2(1) = 0$ or $x = 2$. Now we need one of the original equations in x, y, and z to solve for y. Take $2x + y + z = 0$. (All are equally good.) Then

$$2(2) + y + 1 = 0$$

$$4 + y + 1 = 0$$

$$y + 5 = 0$$

$$y = -5$$

You should check $(2, -5, 1)$ to see that it is a solution to each of the original equations.

Exercises 6-D

I. Find the simultaneous solution to these systems of three equations in three variables.

1. $x - 2y - z = 3$
$2x + y + 3z = 1$
$x - y - z = 5$

2. $x + y + z = 6$
$x + 3z = 4$
$3y - 7z = 2$

3. $x + 4y + 2z = 1$
$3x + 2y - 5z = 25$
$x + y + z = 3$

4. $x + y + 2z = 12$
$x + y - 2z = 0$
$2x - 3y + z = 5$

5. $x + 2y + 3z = 7$
$4y - z = 2$
$z = 3$

6. $x + 2y + z = 5$
$2x - z = 4$
$2y + 3z = 12$

II. Use systems of equations to solve the following problems.

1. A tea merchant has 200 pounds of a mixture made from a Himalayan tea that sells for $7.80 a pound, an Indian tea that sells for $5.20 a pound, and a domestic tea that sells for $3.00 a pound. The mixture contains 4 times as much domestic tea as Himalayan tea. The merchant maintains he must charge $4.27 per pound to make as much money from the mixture as he would selling the teas separately. How many pounds of each tea are in the mixture?

2. A food company is planning to come out with a new instant breakfast food made from three of their regular products—"Tastes Good,"

"Wake-up," and "Hi-pro." Market research has convinced the company that each can should contain 12 ounces of food, 202 calories, and 28 units of protein. "Tastes Good" contains 22 calories and 1 unit of protein per ounce; "Wake-up" contains 20 calories and 1 unit of protein per ounce; and "Hi-pro" contains 15 calories and 3 units of protein per ounce. How should the new mixture be put together?

REVIEW PROBLEMS

1. Give the slope and y intercept for each of the following equations. Then graph the equation.
   (a) $y = -2x + 1$
   (b) $x + 3y - 9 = 0$
   (c) $4y = 12$
2. Give the x intercept and the y intercept for the equation $3x - 2y = 6$. Then graph the equation.
3. Find the slope of the line through $(4, -3)$ and $(6, 2)$.
4. Write an equation for a line that
   (a) has slope $\frac{1}{2}$ and y intercept -1.
   (b) goes through $(2, 2)$ and $(-4, 10)$.
   (c) has x intercept 5 and y intercept -5.
   (d) goes through $(0, 1)$ and is parallel to $x - 2y = 4$.
   (e) has undefined slope and x intercept 3.
   (f) is parallel to the horizontal axis and goes through $(1, -5)$.
5. For each of the following pairs of linear equations, find the common solution if there is one. If there is no solution or if there is an infinite number of solutions, state that.   _parallel lines_   _same line_
   (a) $2x - 3y = 2$
       $x - 5y = -6$
   (b) $7x + 9y = -3$
       $5x - 2y = -19$
   (c) $x + 2y = -1$
       $2x + 4y = 5$
6. Find the simultaneous solution to the following system.

$$3x - 2y + z = 8$$
$$x + 4y - 2z = -2$$
$$5x + y - 3z = 4$$

7. The ninth-grade class sold 950 pounds of sausage and cheese taking in $1978. If the sausage sold for $1.80 a pound and the cheese for $2.20 a pound, how many pounds of sausage and how many pounds of cheese were sold?

8. The price of a pound of hamburger is $1.29. The price of a pound of soybean meal is $0.39. How much hamburger and how much soybean meal should be mixed to get 150 pounds of a mixture that will sell at $0.99 a pound?

9. Property valuation in Centerburg is computed at 35 percent of the market value of the property. Property taxes are figured on the valuation. Write an equation that expresses the tax money T generated by a 5.5 mills tax ($0.0055 per dollar of valuation) on a house with market value V. Use your equation to compute the property tax on a house with a market value of $75,000.

CHAPTER 7

LINEAR INEQUALITIES IN TWO VARIABLES

7.1   SOLUTIONS OF A LINEAR INEQUALITY

In Section 6.2 we examined a problem concerning the sale of refreshments by a committee. Consider now the following related problem.

A committee selling refreshments at a community function charges 50¢ for a cup of coffee and $1 for a stein of beer. The members hope to collect more than $200 through the sales of these beverages. How much coffee and beer could they sell and reach their goal?

The first thing to observe about this problem is that there are many solutions. For example, they could sell 150 coffees and 150 beers, collecting $225; or they could sell 100 coffees and 160 beers collecting $210; or they could sell 400 coffees and 2 beers collecting $202. Obviously, there are many other solutions, as well as many possibilities which are not solutions.

If we let x represent the number of cups of coffee sold and y the number of steins of beer sold, then the solutions to this problem must be pairs of whole numbers satisfying the following conditions.

$$x \geq 0$$

$$y \geq 0$$

$$0.50x + 1.00y > 200$$

The first and second conditions reflect the fact that x and y cannot be negative numbers. The third condition says essentially that the money collected for coffee plus the money collected for beer must exceed $200.

Figure 7.1 shows the line $0.50x + 1.00y = 200$ and the axes; in this section we explain why the shaded region contains all the solutions to this problem. If you select any pair (x, y) of whole numbers from the shaded region, it will satisfy the conditions; any pair selected outside the shaded region will not satisfy the conditons. Try the points indicated on Figure 7.1 and others of your choice.

Examine the graph in Figure 7.1 further by considering some particular values of x, the number of coffees sold. If $x = 100$, then $50 will be collected from the sale of coffee, and 150 beers must be sold to net $200. For $x = 100$ and any value of y greater than 150, the profit will exceed $200; thus, the part of the line $x = 100$ that has y coordinates greater than 150 will lie within the graph of the solutions to the problems. You should make a comparable analysis of the graph when 200 cups of coffee are sold, and when 250 cups of coffee are sold. What parts of the lines $x = 200$ and $x = 250$ lie on the graph?

Inequalities in two variables apply in many situations; in Section

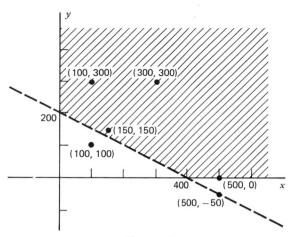

Figure 7.1

7.3 we examine some of these applications. However, we must first study the solutions of linear inequalities like $0.50x + 1.00y > 200$.

In Chapter 2 (and again in Chapter 4) we studied inequalities in conjunction with the trichotomy principle, which stated that for any two numbers a and b exactly one of the following is true: $a < b$, $a = b$, or $a > b$. When this principle is applied to the expression $3x + 2y$ and the number 12, it yields

> For any pair of values of x and y,
> exactly one of the following is true:

$$3x + 2y < 12$$

$$3x + 2y = 12$$

$$3x + 2y > 12$$

To solve a linear inequality in two variables, we must find all the pairs of values of x and y that make an inequality like $3x + 2y < 12$ or $3x + 2y > 12$ true. In general, this can be done graphically and can be related to the graphic solution of the equation $3x + 2y = 12$.

In Chapter 5 we examined the solutions to the equation $3x + 2y = 12$ and found their graph to be a line as shown in Figure 7.2. Notice that this line (or any line) divides the coordinate plane into three pieces: the line itself and two "half-planes." In Figure 7.2 one half plane is marked with dots (▦) and the other with lines (⧄).

The solutions to the equation $3x + 2y = 12$ all lie on the line; the solutions to the inequality $3x + 2y < 12$ are the pairs of values of x and y in one half-plane, and the solutions to $3x + 2y > 12$ are the

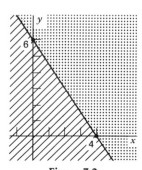

Figure 7.2

pairs of values in the other half-plane. To see this, rewrite the equation $3x + 2y = 12$ as $y = -\frac{3}{2}x + 6$, and take a particular value of x, say $x = 2$. The y value, $y = -\frac{3}{2}(2) + 6 = 3$, identifies a point on the line that is the graph of the equation. A y value greater than 3 identifies a point above the point $(2, 3)$, and a y value less than 3 identifies a point below the point $(2, 3)$.

The graph of $y = -\frac{3}{2}x + 6$ is a line.

The graph of $y > -\frac{3}{2}x + 6$ is the region above the line (▦).

The graph of $y < -\frac{3}{2}x + 6$ is the region below the line (▨).

(Now would be a good time to do Part I of Exercises 7-A.)

Remember that the equation $y = -\frac{3}{2}x + 6$ is equivalent to the equation we started with, $3x + 2y = 12$. Also, the inequality $y > -\frac{3}{2}x + 6$ is equivalent to $3x + 2y > 12$, and $y < -\frac{3}{2}x + 6$ is equivalent to $3x + 2y < 12$. If you graph the following four points, you should see that they lie in the half-plane below the line $3x + 2y = 12$ (the dotted area in Figure 7.2): $(0, 0)$, $(-1, 5)$, $(-1, -1)$, $(2, 2)$. Check to see that these points satisfy the inequality $3x + 2y < 12$. If you graph these four points, you should see that they lie in the half-plane above the line $3x + 2y = 12$ (the squiggled area in Figure 7.2): $(1, 6)$, $(4, 4)$, $(0, 8)$, $(8, 0)$. Check to see that these points satisfy the inequality $3x + 2y > 12$.

For any linear inequality $ax + by > c$ or $ax + by < c$ where a, b, and c are numbers, a procedure for finding the solutions to the inequality is first to graph the line $ax + by = c$, then find out which of the two half-planes determined by the line is associated with the inequality. One convenient way to do this is to choose a point that is *not* on the line. If that point satisfies the inequality, its half-plane is the set of solutions; if the point does not satisfy the inequality, then the other half-plane is the solution set. Consider the following examples.

EXAMPLE 1   Graph the solutions of $2x + 5y < 10$. Examine Figure 7.3.

1. Find the graph of $2x + 5y = 10$. Make a nonsolid line to indicate that the line is not part of the solution to the inequality $2x + 5y < 10$.

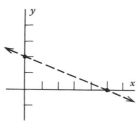

Figure 7.3

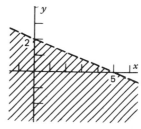

Figure 7.4

2. $(0, 1)$ is a point not on the line. Evaluate $2x + 5y$ at $(0, 1)$.
$$2 \cdot 0 + 5 \cdot 1 = 5 < 10$$

3. Therefore $(0, 1)$ satisfies $2x + 5y < 10$. The half-plane containing $(0, 1)$ represents the solutions of $2x + 5y < 10$. (See Figure 7.4. Note that the line itself is not part of the solutions to the inequality.)

EXAMPLE 2   Graph the solutions of $2x - 3y < 6$.

1. Find the graph of $2x - 3y = 6$. (See Figure 7.5.)

2. $(0, 0)$ is a point *not* on the line and $2 \cdot 0 + 3 \cdot 0 = 0 < 6$. Therefore, $(0, 0)$ satisfies $2x + 3y < 6$ and is on the desired graph.

3. Shade the half-plane containing $(0, 0)$. (See Figure 7.6.)

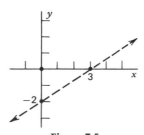

Figure 7.5

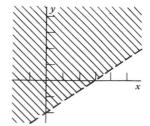

Figure 7.6

EXAMPLE 3   Graph the solutions of $-x - 2y \leq 2$.

1. The solutions of this inequality will include solutions to $-x - 2y = 2$ and $-x - 2y < 2$. (Draw a solid line to indicate that this *is* part of the solution to the inequality. See Figure 7.7.)

2. $(0, 0)$ is a point *not* on the line and $-0 - 2(0) = 0 < 2$. Therefore, $(0, 0)$ satisfies $-x - 2y < 2$ and is on the desired graph.

3. Shade the half-plane containing $(0, 0)$. (See Figure 7.8.)

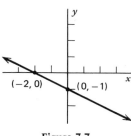

Figure 7.7

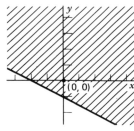

Figure 7.8

A common error made in graphing solutions to linear inequalities is graphing the "upper" half-plane whenever the inequality symbol is ">" or "≥" and the lower half-plane whenever the symbol is "<" or "≤". Notice that this strategy is not always correct; in Examples 2 and 3 above, it would lead to an error. The best way to determine which half-plane is associated with an inequality is to select and test a convenient point in either half-plane. When $(0, 0)$ is not on the line, it is a good choice for the point to test; when it is on the line another point must be selected.

Exercises 7-A

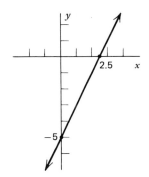

Figure 7.9

I. Figure 7.9 shows the graph of $2x - y = 5$.

On Figure 7.9, graph the following equations.

1. $2x - y = 1$ 4. $2x - y = -3$
2. $2x - y = 3$ 5. $2x - y = 6$
3. $2x - y = 0$ 6. $2x - y = 9$
7. How are these lines related to $2x - y = 5$? How do the last two differ from the first four? Explain.

II. Figure 7.10 shows the graph of $y = 3x - 2$.

For each of the following values of x, show on the plane all points having the given x coordinate and satisfying $y > 3x - 2$.

1. $x = 0$ 4. $x = \dfrac{1}{3}$

2. $x = 2$ 5. $x = -1$

3. $x = -2$

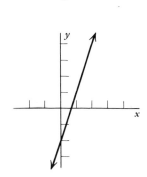

Figure 7.10

III. Graph the solutions of the following inequalities.

1. $3x + y < 5$ 6. $4x + 3y \leq -12$
2. $x + 3y \leq 5$ 7. $4x + 3y \geq 12$
3. $x - y < 0$ 8. $5x - y > 2$
4. $2x - y \leq 3$ 9. $25x - 10y > 5$
5. $x - 2y > 1$ 10. $x + 2y \geq 0$

11. $x + y - 6 > 0$ **16.** $3x \geq 5$

12. $6x - 3y + 1 \geq 7$ **17.** $y < 4$

13. $x \leq 4 + 2y$ **18.** $4x < -3y$

14. $3x + 6 \geq 2y$ **19.** $x \geq 0$

15. $27x \geq 80 + y$ **20.** $2x - 3y > 0$

7.2 SIMULTANEOUS SOLUTIONS OF LINEAR INEQUALITIES

The problem examined at the beginning of this chapter involved not one but three linear inequalities: $x \geq 0$, $y \geq 0$, and $0.50x + 1.00y > 200$. The solutions to that problem are those pairs of values for x and y that satisfy all three of the conditions, that is, the simultaneous solutions to the three inequalities. Graphically, we can describe the solution as those points that are in the region where the solutions of all three inequalities overlap. Figures 7.11 to 7.15 show the graphs of $x \geq 0$, $y \geq 0$, and $0.50x + 1.00y > 200$ separately and on the same

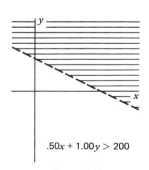

Figure 7.11

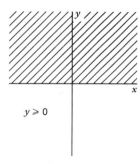

Figure 7.12

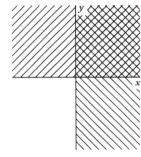

Figure 7.13

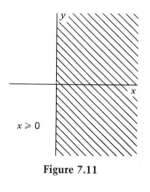

Figure 7.14

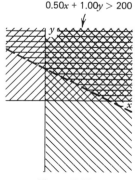

Figure 7.15

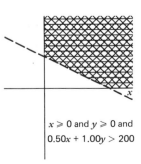

Figure 7.16

coordinate plane; compare the final figure—Figure 7.16—with Figure 7.1. The region shaded to indicate solutions of all three inequalities is the region containing the solutions to the problem.

In general, the simultaneous solutions to two or more inequalities are found by determining the region of overlap for the solutions of the separate inequalities. Consider the pair of equations $x + y = 0$ and $y - 2x = 0$. These lines are graphed in Figure 7.17. In this figure, broken lines ($\underline{\underline{\diagdown}}$) are used to indicate the region containing solutions to $x + y > 0$ and crosses ($\cdot{}^{+}{}_{+}\cdot$) indicate the region containing solutions to $y - 2x > 0$. The simultaneous solutions to $x + y > 0$ and $y - 2x > 0$ are in the regions where broken lines and crosses both appear.

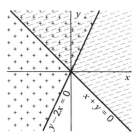

Figure 7.17

Figure 7.17 shows that the lines $y - 2x = 0$ and $x + y = 0$ actually divide the plane into four regions. Each of these regions represents the simultaneous solutions of a pair of inequalities as follows.

Simultaneous Solutions to:	*Indicated by:*
$x + y > 0$ $y - 2x > 0$	Overlap of ($\underline{\underline{\diagdown}}$) and ($\cdot{}^{+}{}_{+}\cdot$).
$x + y > 0$ $y - 2x < 0$	($\underline{\underline{\diagdown}}$) present but no ($\cdot{}^{+}{}_{+}\cdot$).
$x + y < 0$ $y - 2x > 0$	($\cdot{}^{+}{}_{+}\cdot$) present but no ($\underline{\underline{\diagdown}}$).
$x + y < 0$ $y - 2x < 0$	No ($\underline{\underline{\diagdown}}$) or ($\cdot{}^{+}{}_{+}\cdot$).

In light of this analysis, consider the following examples.

EXAMPLE 1   Find the simultaneous solution to

$$x + 2y \geq 4$$
$$2x - 3y < 1$$

1. Begin by graphing the solutions to $x + 2y \geq 4$ (Figure 7.18).

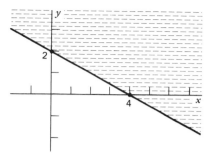

Figure 7.18

2. Then graph $2x - 3y < 1$ on the same axes (Figure 7.19).

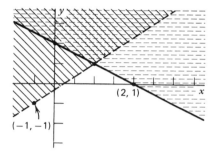

Figure 7.19

The region of overlap is the graph of the simultaneous solutions (Figure 7.20).

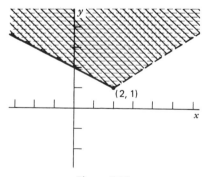

Figure 7.20

EXAMPLE 2   Find the simultaneous solutions to

$$2x + 3y > 6$$
$$2x + 3y < 12$$

1. Graph the solutions to $2x + 3y > 6$ (Figure 7.21).

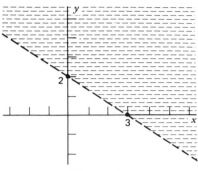

Figure 7.21

2. On the same coordinate system graph $2x + 3y < 12$ (Figure 7.22).

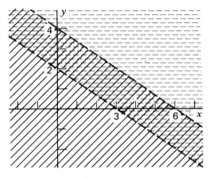

Figure 7.22

The region of overlap is the graph of the simultaneous solutions (Figure 7.23).

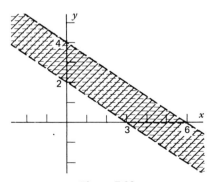

Figure 7.23

Examine the last example carefully. Can you see that the following pairs of simultaneous inequalities have the graphs indicated in Figures 7.24 and 7.25?

$2x + 3y > 6$
$2x + 3y > 12$

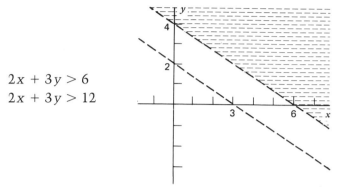

Figure 7.24

$2x + 3y < 6$
$2x + 3y < 12$

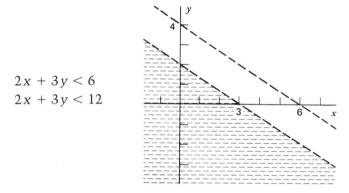

Figure 7.25

$2x + 3y < 6$
$2x + 3y > 12$ No solutions!

Exercises 7-B

I. Graph the simultaneous solutions for the following pairs of inequalities.

1. $x + y < 7$
   $x - y > 3$

2. $2x + y \leq 5$
   $y < 3$

3. $3x + 2y \leq 19$
   $2x + 2y \leq 16$

4. $2x + 3y > 13$
   $x + 2y < 8$

5. $2x - 3y \leq 10$
   $3x - 4y \geq 11$

6. $5x + 3y > 0$
   $8x + 2y < 0$

7. $4x + 3y \geq 12$
$4x + 3y \leq 24$

8. $4x + y < 1$
$6x + 2y \geq -6$

9. $x + y < 3$
$2x + 2y > 8$

10. $x - 3y < 3$
$2x - y > 1$

II. Graph the simultaneous solutions for each family of inequalities.

1. $x > 0$
$y < 5$
$x + y > 4$

2. $x + y < 10$
$x - y > 2$
$y > 0$

3. $2x + y < 5$
$x + 2y > 1$
$x - y < 5$

4. $x + 2y + 4 > 0$
$x + 2y - 4 < 0$
$x \geq -4$
$y \geq -2$

5. $x + y < 1$
$x + y > -1$
$x - y < 1$
$x - y > -1$

7.3   APPLICATIONS OF LINEAR INEQUALITIES

Consider this problem.

A bakery can operate its mixer and oven 24 hours a day manufacturing two kinds of cookies in large batches. Each batch of Crunchies requires use of the mixer for 3 hours and use of the oven for 2 hours. Each batch of Crispies requires the mixer for $1\frac{1}{2}$ hours and the oven for 3 hours. How many batches of Crunchies and Crispies can the bakery plan to make in a day?

The key to solving this problem is to recognize that the factors determining the solution are time limitations, that is, neither the mixer nor the oven can be used for more than 24 hours per day. With this in mind, let x be the number of batches of Crunchies and let y be the number of batches of Crispies.

Consider the use of the mixer to determine

$$3x + 1\frac{1}{2}y \leq 24$$

Then consider the use of the oven to determine

$$2x + 3y \leq 24$$

It is also true that

$$x \geq 0 \quad \text{and} \quad y \geq 0$$

The simultaneous solution to these inequalities is illustrated by the cross-hatching in Figure 7.26. The region of overlap is referred to as

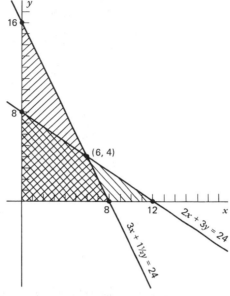

Figure 7.26

the "feasible set." Notice that the feasible possibilities include 6 batches of Crunchies and 4 batches of Crispies, 1 batch of Crunchies and 7 batches of Crispies, and many other possibilities.

Consider a second problem.

A manufacturer plans to produce a new breakfast cereal from two of its generic products, Pro and Super. Each ounce of Pro contains 4 units of vitamin P and 3 units of vitamin Q. An ounce of Super contains 1 unit of vitamin P and 2 units of vitamin Q. How can they mix Pro and Super to obtain a package of 6 to 8 ounces that contains at least 10 units of vitamin P and 15 units of Q.

Again, the conditions of the problem are conditions of inequality. If x is the number of ounces of Pro and y is the number of ounces of Super, we arrive at the following inequalities.

$x \geq 0$

$y \geq 0$

$x + y \geq 6$ \qquad (Package can contain at least 6 ounces)

$6 \leq x + y \leq 8$

$$x + y \leq 8 \qquad \text{(Package can contain 8 ounces or less)}$$
$$4x + y \geq 10 \qquad \text{(Amount of Vitamin P)}$$
$$3x + 2y \geq 15 \qquad \text{(Amount of vitamin Q)}$$

To find the feasible set, graph these inequalities.

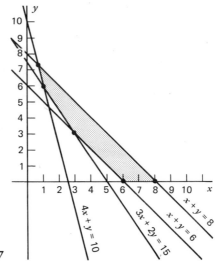

Figure 7.27

The feasible set is shaded heavily in Figure 7.27.

The problems in Exercises 7-C contain other applications of linear inequalities to business situations.

Exercises 7-C

Solve the following problems.

1. The Winters want to buy a gift for both Tom and Jerry. They don't want to spend more than $50 altogether, and they feel that Tom's gift should be at least $5 more valuable than Jerry's. Neither gift should cost less than $10. Sketch the region showing the amounts they can pay for each gift.

2. A caterer, Mrs. Jones, has been contracted to supply two kinds of sandwiches at a party for 240 people. She needs to allow at least $1\frac{1}{2}$ sandwiches per person and knows from experience that the ham sandwiches will be about twice as popular as the corned beef. Mrs. Jones charges the party host $1 for each ham sandwich and $1.50 for each corned beef sandwich; the total cost of the sandwiches should not exceed $420. Graph the feasible set.

7.4 APPLICATIONS TO COST AND PROFIT

Return again to the problem of mixing Pro and Super cereals (on page 115). In our examination of that problem we found all possible ways of mixing Super and Pro to meet certain standards and found that there were many ways to do it. From the manufacturer's point of view, however, we left out some very important considerations, namely those concerned with money! Reconsider now the problem with the added condition that the manufacturer wants to produce the mixture as inexpensively as possible.

A manufacturer plans to produce a new breakfast cereal from two of its generic products, Pro and Super. Each ounce of Pro contains 4 units of vitamin P and 3 units of vitamin Q. An ounce of Super contains 1 unit of vitamin P and 2 units of vitamin Q. It costs the company 10¢ to produce an ounce of Pro and 7¢ to produce an ounce of Super. How much of each should the company mix to produce a box of cereal weighing between 6 and 8 ounces at the lowest cost to the company?

Consider first the cost of a box of cereal. If the box contains x ounces of Pro and y ounces of Super, the cost of producing it is $10x + 7y$ cents. Consider the different possibilities for C—the cost of production. Each of these possibilities gives us a different equation of the form $10x + 7y = C$. Each of these equations has a line for its graph; the lines, called cost lines, have the same slope and thus are parallel, as shown in Figure 7.28 on the following page.

To find the mixture of Pro and Super that meets all the conditions *and* can be produced at lowest cost, we can examine these cost lines in conjunction with the feasible set and ask, "What is the lowest cost associated with a line that meets the feasible set?" "What mixture(s) is(are) on that line?" By drawing several of the cost lines on Figure 7.29, we see that the lowest cost line that touches the feasible set is associated with a cost of 51¢, and this line touches the feasible set at exactly one point—(3, 3). Thus, the most economical decision for the manufacturer is to mix 3 ounces of Pro and 3 ounces of Super in each package.

What would happen if the cost of Pro went up to 11¢ an ounce, and the cost of Super went down to 4¢ an ounce? The feasible set would remain the same, but the cost lines would become the family $11x + 4y = C$. The cost of producing a mixture of 3 ounces Pro and 3 ounces Super would be $11 \cdot 33 + 4 \cdot 3 - 45$¢. However this would not be the most economical choice; mixing 1 ounce of Pro with 6 ounces of Super is feasible and would only cost $11 \cdot 1 + 4 \cdot 6 = 35$¢. Draw some of the cost lines $11x + 4y = C$ on your graph of the feasible set and observe that this is the best solution.

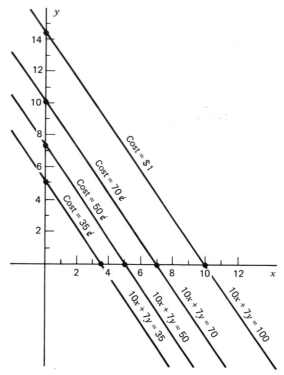

Figure 7.28

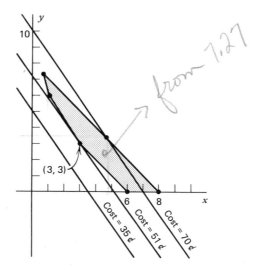

Figure 7.29

In each of the cost situations described here, the minimum cost to the manufacturer is realized at a vertex of the feasible set. This does not happen by accident, In fact, the solution to a problem like this always lies on a vertex of the feasible set. Consequently, we can restrict our attention to the vertices of the feasible set when trying to minimize an expression for cost or time spent, or maximize an expression for profit. To see why the solution occurs on a vertex of the feasible set, consider the feasible set in Figure 7.30. Now place your straightedge on cost line 1 and slide it toward the origin in such a way that it remains parallel to the cost line; the last point of the feasible set that the straight edge touches will represent the minimum cost; this occurs at vertex A. Now place your straightedge on profit line 2 and slide it; do you see that the highest profit would occur at vertex D, while the lowest profit would be realized at vertex B?

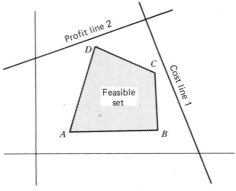

Figure 7.30

Before turning to Exercises 7-D, examine another application of this procedure.

A bakery sells cookies and cupcakes. Each batch of cookies requires 40 minutes in the mixer, 20 minutes in the oven, and 20 minutes with the icer. Each batch of cupcakes requires 20 minutes in the mixer, 40 minutes in the oven, and 30 minutes with the icer. Because of startup and cleanup time, the mixer is operated 480 minutes, the oven 420 minutes, and the icer 320 minutes per day. The profit is $10 per batch of cookies and $16 per batch of cupcakes. How many batches of each should the bakery produce to maximize profit?

To begin work on this problem, we translate each condition into an inequality using x as the number of batches of cookies and y as the number of batches of cupcakes.

$x \geq 0$

$y \geq 0$

$40x + 20y \leq 480$ (Minutes the mixer is used)

$20x + 40y \leq 420$ (Minutes the oven is used)

$20x + 30y \leq 320$ (Minutes the icer is available)

Now graph the feasible set and find its vertices. Examine Figure 7.31. The vertices of the feasible set can be found by determining the points of intersection of the lines forming the boundary of the set.

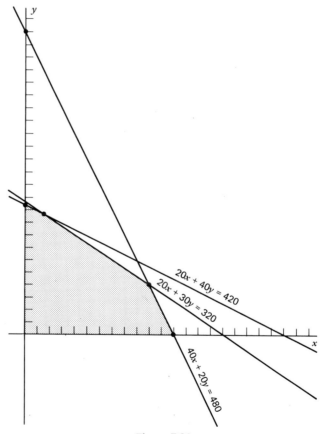

Figure 7.31

$(0, 10.5)$ Solution to $\begin{cases} x = 0 \\ 20x + 40y = 420 \end{cases}$

$(1, 10)$ Solution to $\begin{cases} 20x + 30y = 320 \\ 20x + 40y = 420 \end{cases}$

$$(10, 4) \qquad \text{Solution to} \quad \begin{cases} 20x + 30y = 320 \\ 40x + 20y = 480 \end{cases}$$

$$(12, 0) \qquad \text{Solution to} \quad \begin{cases} 40x + 20y = 480 \\ y = 0 \end{cases}$$

Now we find the profit at each of these vertices. The profit is given by $10x + 16y$.

Vertex	Profit
$(0, 10.5) \rightarrow$	$10 \cdot 0 + 16 \cdot 10.5 = 168$
$(1, 10) \rightarrow$	$10 \cdot 1 + 16 \cdot 10 = 170$
$(10, 4) \rightarrow$	$10 \cdot 10 + 16 \cdot 4 = 164$
$(12, 0) \rightarrow$	$10 \cdot 12 + 16 \cdot 0 = 120$

Thus, the profit will be maximized if the bakery makes 1 batch of cookies and 10 batches of cupcakes each day. In this case the profit is $170.

Exercises 7-D

I. Use your straightedge to determine the vertex of the feasible region at which each of the following occurs. See Figure 7.32.

1. $x + 8y$ is minimized.
2. $x + 8y$ is maximized.
3. $x + y$ is minimized.
4. $x + y$ is maximized.
5. $x + 6y$ is minimized.
6. $x + 6y$ is maximized.
7. $y - 2x$ is minimized.
8. $y - 2x$ is maximized.

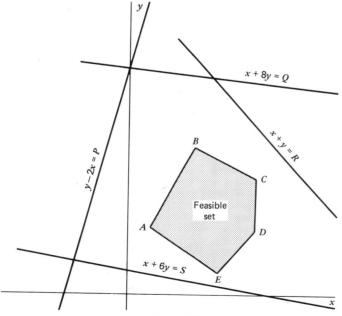

Figure 7.32

II. In each of the following problems graph the feasible region and find the points where the indicated expression is minimized or maximized.

1. Feasible region:

$$x \geq 0$$
$$y \geq 0$$
$$x + y \leq 5$$
$$2x + 3y \leq 13$$

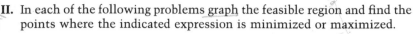

Maximize $3x + 4y$.

2. Feasible region:

$$x \geq 0$$
$$y \geq 0$$
$$y \geq x$$
$$y \leq 10$$
$$x + y \leq 12$$

Maximize $2x - y$.

3. Feasible region:

$$2y + x \leq 6$$
$$x + y \geq 2$$
$$x \geq 0$$
$$y \leq 2x + 2$$

Maximize $2y - x$.

4. Feasible region:

$$6x + y \geq 30$$
$$10x + 5y \geq 70$$
$$8x + 9y \geq 66$$
$$x \geq 0$$
$$y \geq 0$$

Minimize $5x + 4y$.

5. Feasible region:

$$3x + 2y \geq 24$$
$$2x + y \geq 14$$
$$y - x \leq 2$$
$$x \geq 0$$
$$y \geq 0$$

Minimize $x + y$.

III. Solve the following problems.

1. A real estate developer is planning to build an apartment complex on a plot of land. Zoning regulations require that no more than 20 apartments be built on the plot and that the number of one-bedroom apartments be at least $\frac{2}{3}$ as large as the number of two-bedroom apartments. How many of each type (one and two bedroom) should the developer build to maximize his rental income if he can charge

$380 for the one-bedroom apartments and $425 for the two-bedroom apartments?

2. A wholesale salad maker sells two kinds of salad to delicatessens. Each 10-gallon batch of potato salad requires 40 minutes of chopping, 25 minutes of mixing, and 15 minutes of packaging. Each batch of macaroni salad requires 15 minutes of chopping, 10 minutes of mixing, and 10 minutes of packaging. Because of work schedules, all chopping must be done in 480 minutes, mixing in 305 minutes, and packaging in 250 minutes. Profit on potato salad is $25 per batch and on macaroni salad is $17 per batch. Determine the number of gallons of potato salad and the number of gallons of macaroni salad per day that will yield a maximum profit for the wholesaler.

3. A large grocery store carries "fancy" and "generic" canned peaches. Based on market research, the manager has decided no more than 16 cases of canned peaches can be sold per day and that they can sell at least 3 more cases of generic than fancy peaches each day. Each case of generic peaches requires 18 square inches of shelf, and each case of fancy peaches requires 27 square inches of shelf. The manager has allotted a maximum of 324 square inches of shelf space for peaches. If the profit on a case of fancy peaches is $1.20 and the profit on a case of generic peaches is $0.72, how many cases of each should be shelved?

REVIEW PROBLEMS

1. Graph the following inequalities.
   (a) $3x > y$
   (b) $x + y \leq 5$
   (c) $2x - 8 > 3y + 10$
   (d) $5x - 3y \leq 15$

2. Find the simultaneous solutions to each system of inequalities.
   (a) $x + y > 6$
       $x - y < 2$
   (b) $x \geq 5$
       $2x - y \leq 8$
   (c) $4x + 5y \leq 20$
       $4x - 5y \leq 20$
   (d) $2x + 4y > 9$
       $x + 2y < 3$

3. Graph the simultaneous solutions for the following family of inequalities.
$$x \geq 0$$
$$y \geq 0$$
$$x + y \leq 12$$
$$5x + y \geq 24$$
$$x - y \leq 6$$

4. Twenty-eight violinists, 13 flutists, and 10 cellists have joined together and hired an agent to help them find work playing at wedding receptions

and on other occasions. The agent advertises small ensembles consisting of 2 violins, 2 flutes, and 1 cello and large ensembles consisting of 7 violins, 3 flutes, and 2 cellos. Letting x represent the number of small ensembles and y the number of large ensembles, sketch the feasible region showing the number of ensembles that can play at any one time.

CHAPTER 8

POLYNOMIAL ARITHMETIC AND FACTORING

8.1 RECOGNIZING POLYNOMIALS

There is a type of algebraic expression that arises so often in problems that you need to develop skill in computing with it. Here are some examples.

1. Concert tickets cost $10 apiece at the door. However, groups can make arrangements to purchase tickets in advance at a special price. The theater deducts from the price of a ticket 5¢ for each member of the group. Thus, if a group has N persons, each ticket costs $10 - 0.05N$ and the total cost for the group is $N(10 - 0.05N)$ or $10N - 0.05N^2$.

2. Sturdy Box Company has one line of boxes that have square bases with the height of the boxes twice the base length. If a box has base length b, then we can write an expression that describes the volume of the box. Since the area of the base is b^2 and the height $2b$, the expression describing the volume is $(b^2)(2b)$ or $2b^3$.

Expressions like $10N - 0.05N^2$ and $2b^3$ are called **polynomials.** In the polynomial $10N - 0.05N^2$ the number 10 is the **coefficient** of N and -0.05 is the coefficient of N^2. Here are some other polynomials.

$$3x - 7$$

$$\frac{1}{2}z^3 - z + 6$$

$$14t^4 - 49t$$

$$5 - x^3 + 4x$$

A polynomial is a sum; one term of the sum may be a number and the other terms may consist of a number times a power of the variable. To write a general expression that indicates what a polynomial looks like, we might represent the variable by the letter x and assume the powers of x that occur are $x, x^2, x^3, \ldots, x^n$. It is convention to denote the numerical coefficient of x by a_1, the coefficient of x^2 by a_2, the coefficient of x^3 by a_3, and so on. Since one term in a polynomial may be a number without the variable, we denote that number by a_0. Using these symbols then, a polynomial is an expression that can be arranged to look like

$$a_n x^n + \cdots + a_2 x^2 + a_1 x + a_0$$

The numbers $1, 2, \ldots, n$ on the coefficients $a_1, a_2, \ldots, a_n$ are called **subscripts.** The subscript indicates which power of the variable the coefficient belongs to and does not affect the value of the coefficient.

Polynomials can be written in more than one variable: $x^2 - xy +$

[handwritten margin note: div. by a variable is not a polynomial]

y^2, $3x^3 + x^2y - 7y^2$, $x + y - 3x^2$. So many of the expressions that arise in problems are polynomials that students may get the idea that all algebraic expressions are polynomials. This is not the case. Here are some examples of nonpolynomials.

$$\frac{4}{x} - 2 \qquad 6\sqrt{w} + 1 \qquad x^2 - \frac{x-1}{x^2+1} + 3 \qquad \frac{x}{y}$$

[handwritten margin notes: $w^{\frac{1}{2}}$ (pointing to $\sqrt{w}$); rational exp. (bracketing last two)]

8.2 ADDING AND SUBTRACTING POLYNOMIALS

Polynomials, like numbers, can be combined by addition and subtraction. In Chapter 2 the distributive property and combining like terms were discussed; you practiced adding and subtracting polynomials at that time. Any two polynomials can be added and any polynomial can be subtracted from another. The following examples should remind you of these procedures.

$$(x^2 + 7x - 1) + (-3x^2 + 6x) = x^2 + 7x - 1 - 3x^2 + 6x$$
$$= -2x^2 + 13x - 1$$

$$(x^2 + 7x - 1) - (-3x^2 + 6x) = x^2 + 7x - 1 + 3x^2 - 6x$$
$$= 4x^2 + x - 1$$

$$\left(xy + \frac{1}{2}\right) + \left(3y^2 - xy - \frac{1}{2}\right) = xy + \frac{1}{2} + 3y^2 - xy - \frac{1}{2}$$
$$= 3y^2$$

$$\left(xy + \frac{1}{2}\right) - \left(3y^2 - xy - \frac{1}{2}\right) = xy + \frac{1}{2} - 3y^2 + xy + \frac{1}{2}$$
$$= -3y^2 + 2xy + 1$$

8.3 MULTIPLYING POLYNOMIALS

You have also had some practice multiplying polynomials; indeed, many of the examples and problems on the use of the distributive property involved multiplying polynomials. Now we can extend those ideas. Consider the following computation.

$$(x^2 - 2x + 1)(3x + 4) = (x^2 - 2x + 1)(3x) \qquad \text{(Distributive}$$
$$+ (x^2 - 2x + 1)(4) \qquad \text{property)}$$
$$= (3x^3 - 6x^2 + 3x) \qquad \text{(Distributive}$$
$$+ (4x^2 - 8x + 4) \qquad \text{property again!)}$$
$$= 3x^3 - 2x^2 - 5x + 4 \qquad \text{(Combining}$$
$$\text{like terms)}$$

The distributive property in the first line of the computation has the effect of requiring that each term in the second polynomial, $3x + 4$, multiplies the first polynomial, $x^2 - 2x + 1$. The distributive property in the second line guarantees that each term of the first polynomial multiplies each term of the second polynomial; the third line says that like terms then are combined. Sometimes polynomial multiplication is displayed in a vertical rather than a horizontal format. Study the following computation to see that the result is the same: each term in the second polynomial, $3x + 4$, multiplies each term in the first, $x^2 - 2x + 1$, and then all of these products are combined. The products are arranged to make the additions easy.

$$
\begin{array}{r}
x^2 - 2x + 1 \\
3x + 4 \\
\hline
4x^2 - 8x + 4 \qquad \text{(Multiplying by 4)} \\
3x^3 - 6x^2 + 3x \qquad \text{(Multiplying by } 3x) \\
\hline
3x^3 - 2x^2 - 5x + 4
\end{array}
$$

To compute products of polynomials in more than one variable, we must use the same procedure. The following computation gives the product $(x - 2y)(x^2 + xy - 9y^2)$.

$$
\begin{array}{r}
x^2 + xy - 9y^2 \\
x - 2y \\
\hline
- 2x^2y - 2xy^2 + 18y^3 \\
x^3 + x^2y - 9xy^2 \\
\hline
x^3 - x^2y - 11xy^2 + 18y^3
\end{array}
$$

Notice that like terms are placed in the same column so that they can be easily added.

Some polynomial multiplications occur so often that it helps to remember the pattern of the product. Consider the product of two linear polynomials.

$$(x + 3)(2x - 5) = 2x^2 - 5x + 6x - 15$$
$$= 2x^2 + x - 15$$
$$(-6x + 1)(2x + 3) = -12x^2 - 18x + 2x + 3$$
$$= -12x^2 - 16x + 3$$

An important thing to observe is that the four terms resulting from the multiplication of two linear polynomials in the same variable combine to make three terms. The first term is the product of the two first terms in the linear polynomials, the last term is the product of the two last terms, and the middle term is a sum. The middle term can be represented this way.

[Handwritten margin notes: "Linear polynomial = Highest exponent on variable is 1." "How do we know it's linear?"]

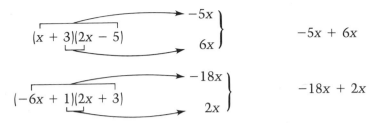

Polynomials consisting of two terms are called **binomials.** The linear polynomials above are binomials. Whenever two binomials consisting of like terms are multiplied together, the pattern identified above emerges. Here are some additional examples.

$$(x^2 + 2x)(3x^2 - x) = 3x^4 - x^3 + 6x^3 - 2x^2$$

$$= 3x^4 + 5x^3 - 2x^2$$

$$(2x - y)(x + 5y) = 2x^2 + 10xy - xy - 5y^2$$

$$= 2x^2 + 9xy - 5y^2$$

Observe again that the middle term is a sum. Sometimes the middle term of the product becomes zero as in this example.

$$(2x - 5)(2x + 5) = 4x^2 - 25$$

Can you explain in what situations the middle term is zero?

Exercises 8-A

I. Find the following products using either a horizontal or a vertical computation.

1. $(x + 3y)(2x - y)$

2. $(x^2 + 5x - 1)(x^2 - 2x + 1)$

3. $(x^2 + 2)(3x^2 + x + 1)$

4. $(2r + 3)(r^2 - r + 2)$

5. $(r^5 - 1)(r^3 + 2r - 1)$

6. $(r - 1)(r^3 + r^2 + r + 1)$

7. $(x + 2y)(x^2 - xy + y^2)$

8. $(x - y + z + 5)(x + y + z - 5)$

9. $2x^2(x - 4)(x^3 + 2x - 1)$

II. Find the following products.

1. $(2x + 5)(3x + 2)$

2. $(-x + 4)(2x - 1)$

3. $(142x + 6)\left(\frac{1}{2}x - 1\right)$

4. $(7 - a)(2 + 3a)$

5. $(x + 2y)(3x - y)$

6. $(x + 7)(x - 7)$

7. $(2x - y)(2x + y)$

8. $(x + y - 1)(x + y + 1)$

9. $(x + 4)(x + 4)$

10. $(b - 3)^2$

11. $(4y - 2)^2$

12. $(2xy^2 - 3y)^2$

13. $(x - y)(x + y)(x^2 - y^2)$

14. $(x + xy)(2y + x)$

III. Simplify the following expressions by doing the indicated multiplication and then combining like terms.

1. $(x + y)^2 - (x - y)^2$
2. $x^2 + y^2 - (x + y)(x - y)$
3. $(2x - y)(x + 2y) - 3xy$
4. $6y^2 - (x + y)(2x - y)$

IV. Find numbers c and d that complete the factoring in the following equations.

1. $(x + 7)(cx + d) = 2x^2 + 20x + 42$
2. $(x + c)(2x + d) = 2x^2 + x - 36$
3. $(cr + d)(3r + 7) = 6r^2 + 11r - 7$
4. $(x + c)(x + d) = x^2 - 25$
5. $(2x + c)(2x + d) = 4x^2 - 25$
6. $(c - r)(d + r) = 12 + 4r - r^2$

V. Solve the following problems.

1. A company makes custom tablecloths and other fine linens. To allow material for trimming and hemming, they always cut 3 inches of extra fabric in each dimension. They charge customers by the number of square inches of fabric cut. Write a formula for
   (a) the number of square inches cut for a square cloth finished at x inches per side.
   (b) the number of square inches cut for a rectangular cloth that is twice as long as it is wide.
2. A concrete walk of uniform width extends around a rectangular lawn having dimensions 20 by 80 feet. If the width of the walk is x feet, give an expression for its area.
3. A circular picture frame with a radius of 5 inches contains a round print with a border. If the print has radius x inches, write an expression for the area of the border.

8.4   DIVIDING POLYNOMIALS

There are times when we want to divide one polynomial by another. Remember what division means for integers.

$$6 \div 2 = 3 \quad \text{because} \quad 3 \cdot 2 = 6$$

Similarly,

$$(x^2 + 5x + 6) \div (x + 2) = (x + 3)$$

$$\text{because} \quad (x + 3)(x + 2) = x^2 + 5x + 6$$

In discussing the division of polynomials, it helps to have a word for the largest exponent in a polynomial. For polynomials in one variable the largest exponent is called the **degree** of the polynomial. For example, $x^2 + 5x + 6$ is a polynomial of degree 2; a linear polynomial like $3x + 2$ has degree 1. We regard a single number as a polynomial of degree 0. Thus, -6 and 15 are degree 0 polynomials.

The procedure for dividing one polynomial by another is analogous to the procedure for dividing one integer by another. First, we arrange the terms in both polynomials so that the exponents are decreasing. Then, we find the terms of the quotient one at a time, multiply each term by the divisor, and subtract. Analyzing the following example should clarify the procedure.

$$
\begin{array}{r}
3x^2 + 15x - 1 \\
(x - 4)\,\overline{)\,3x^3 + 3x^2 - 61x + 4} \\
\underline{3x^3 - 12x^2} \\
15x^2 - 61x + 4 \\
\underline{15x^2 - 60x} \\
-x + 4 \\
\underline{-x + 4} \\
0
\end{array}
$$

The example above is actually computed in three steps.

1.
$$
\begin{array}{r}
3x^2 \\
x - 4\,\overline{)\,3x^2 + 3x^2 - 61x + 4} \\
\underline{3x^3 - 12x^2} \\
15x^2 - 61x + 4
\end{array}
$$

The first term $3x^2$ in the quotient is chosen because it multiplies $x - 4$ to give a polynomial starting with $3x^3$. Subtracting the second polynomial from the first then gives a polynomial of degree 2 rather than 3.

2.
$$
\begin{array}{r}
3x^2 + 15x \\
x - 4\,\overline{)\,3x^3 + 3x^2 - 61x + 4} \\
\underline{3x^3 - 12x^2} \\
15x^2 - 61x + 4 \\
\underline{15x^2 - 60x} \\
-x + 4
\end{array}
$$

The second term $15x$ is chosen because it multiplies $x - 4$ to give a polynomial starting with $15x^2$. Subtraction now gives a polynomial of degree 1.

3.
$$
\begin{array}{r}
3x^2 + 15x - 1 \\
x - 4\,\overline{)\,3x^2 + 3x^2 - 61x + 4} \\
\underline{3x^3 - 12x^2} \\
15x^2 - 61x + 4 \\
\underline{15x^2 - 60x} \\
-x + 4 \\
\underline{-x + 4} \\
0
\end{array}
$$

The third term -1 is chosen because it multiplies $x - 4$ to give a polynomial starting with $-1x$. This time the remainder is 0.

The above procedure guarantees that $(3x^2 + 15x - 1)(x - 4) = 3x^3 + 3x^2 - 61x + 4$. (Perform the multiplication and check!)

The division process is complete when subtraction yields a polynomial of degree less than the degree of the divisor. This polynomial is called the **remainder.** In the example above the remainder is 0; often (as in the example below) it is not 0. The division procedure is analogous to the example above even if the remainder is not 0. Study the following computation for $(4x^4 + 5x^2 - x + 3) \div (2x^2 + x)$. Notice that the first step is to write $4x^4 + 5x^2 - x + 3$ as $4x^4 + 0x^3 + 5x^2 - x + 3$; this provides a position for the x^3 terms.

$$
\begin{array}{r}
2x^2 - x + 3 \\
2x^2 + x \overline{\smash{\big)}\, 4x^4 + 0x^3 + 5x^2 - x + 3} \\
\underline{4x^4 + 2x^3} \\
-2x^3 + 5x^2 - x + 3 \\
\underline{-2x^3 - x^2} \\
6x^2 - x + 3 \\
\underline{6x^2 + 3x} \\
-4x + 3
\end{array}
$$

The computation displays a quotient of $2x^2 - x + 3$ and a remainder of $-4x + 3$. It demonstrates that $4x^4 + 5x^2 - x + 3 = (2x^2 - x + 3)(2x^2 + x) + (-4x + 3)$. You can perform the multiplication and addition to check this result.

The procedure we used above for dividing one polynomial into another to give a quotient and a remainder is sometimes called the division algorithm. The term **division algorithm** is also used to designate the fact that one polynomial can always be divided into another to give a quotient and a remainder that is either 0 or a polynomial of smaller degree. This division algorithm can be stated in terms of the multiplication statement that corresponds to the division computation.

DIVISION ALGORITHM

If P and D are polynomials, it is always possible to find polynomials Q and R so that

$$P = Q \cdot D + R$$

and R is either 0 or is a polynomial with degree less than the degree of D.

Exercises 8-B

Perform the division in each of the following, giving both the quotient and remainder. Then write the multiplication statement that is guaranteed by the division computation.

1. $(2x^2 - 7x - 4) \div (x - 4)$

2. $(-3x^2 + 7x - 2) \div (3x - 1)$

3. $(2x^4 - x^3 + 7x^2 - 3x + 3) \div (x^2 + 3)$

4. $(2x^3 + x - 5) \div (x^2 - 2x + 7)$

5. $(x^3 - 1) \div (x - 1)$

6. $(x^3 + 2x^2y + 3xy^2 + 4y^3) \div (x^2 + xy + 2y^2)$

7. $(x^3 - 1) \div x^5$

8. $x^5 \div (x^3 - 1)$

9. $(x^4 - 13x^2 + 36) \div (x^2 - 5x + 6)$

10. $(x^8 - x^7 + x^6) \div (x^4 + x^3 - x^2)$

8.5 FACTORING

Any natural number can be written as a product of natural numbers. For example, $12 = 6 \cdot 2$ or $12 = 3 \cdot 4$; $7 = 1 \cdot 7$. A natural number n greater than 1 is called **composite** if it can be written as a product of two natural numbers both different from n. Thus, 12 is a composite number, and so is 10 because $10 = 2 \cdot 5$. Natural numbers greater than 1 that are not composite are called **prime.** If n is prime, it cannot be written as a product of two natural numbers both different from n. Indeed, n is prime if it can only be written as the product of 1 and n. Thus, 2 and 3 and 5 and 7 are prime numbers; so are 31 and 79.

When an expression is written as a product, we say it has been **factored.** An important property of natural numbers is that each composite number greater than 1 can be factored into a product of prime numbers. For example,

$$105 = 3 \cdot 5 \cdot 7$$

$$8 = 2 \cdot 2 \cdot 2$$

$$18 = 2 \cdot 3 \cdot 3$$

$$1001 = 7 \cdot 11 \cdot 13$$

A number that is written as a product of prime numbers is said to be **factored completely.**

We can also talk about factoring polynomials—that is, writing polynomials as products of polynomials of lower degree. In factoring polynomials we restrict ourselves to polynomials with integer coefficients. Check the following statements.

$$x^3 - 2x^2 + 7x = x \cdot (x^2 - 2x + 7)$$
$$2x^2 + x - 3 = (x - 1)(2x + 3)$$

In these examples, x and $x^2 - 2x + 7$ are called **factors** of $x^3 - 2x^2 + 7x$; similarly, $x - 1$ and $2x + 3$ are called factors of $2x^2 + x - 3$. A polynomial that cannot be written as a product of two factors of smaller, positive degree is said to be **irreducible.** Thus, $x - 3$ is irreducible; so is $x^2 + x + 1$. Irreducible polynomials are analogous to prime numbers. Indeed, every polynomial is either irreducible itself or can be expressed as a product of irreducible polynomials. A polynomial that is written as a product of irreducible polynomials is said to be **factored completely.**

Finding the factors of a polynomial (or convincing yourself that it is irreducible) can be a difficult business. There are not step-by-step procedures that will always yield the factors of any polynomial. In fact, many polynomials are irreducible and cannot be factored further. Many that can be factored are difficult to factor. We restrict our attention to a few types of polynomials that fit special patterns. Even then, considerable trial and error may be required. Fortunately, you can always check the product to be sure it does equal the original polynomial.

In the next few paragraphs we consider two particular patterns: one is a consequence of the distributive property and the other describes certain polynomials of degree 2. Other patterns are included in the exercises.

(1) DISTRIBUTIVE PROPERTY

Remember that $ab + ac = a(b + c)$. This fact provides a way of factoring polynomials in which each term contains the same factor.

$$2x^2 + x = x(2x + 1)$$
$$4x + 8y + 6z = 2(2x + 4y + 3z)$$
$$4x^2y - 8xy^2 = 4xy(x - 2y)$$
$$(x - y) + 3(x - y)z^2 = (x - y)(1 + 3z^2)$$

A rule of thumb in factoring a polynomial is first to identify the factor that is common to each term and then to write the polynomial as that factor times another polynomial. In short, find the common factor first. For example, if the polynomial to be factored is $6x^2y + 42xy + 36y$, we identify the common factor $6y$ and write

$6x^2y + 42xy + 36y = 6y(x^2 + 7x + 6)$. The next step is to see if $x^2 + 7x + 6$ can be factored. Read on!

(2) POLYNOMIALS OF DEGREE 2

Polynomials of degree 2 are often called **quadratic** polynomials. Some quadratics can be written as the product of two linear polynomials. If we take two linear polynomials $ax + b$ and $cx + d$, where a, b, c, and d are integers and x is the variable, we can compute their product.

$$(ax + b)(cx + d) = acx^2 + (ad + bc)x + bd$$

This equation shows the relationship between the coefficients of the factors $ax + b$ and $cx + d$ and the coefficients of the quadratic that is their product. In the simplest case, a and c are both equal to 1 and this equation becomes

$$(x + b)(x + d) = x^2 + (b + d)x + bd$$

We begin with an example of this case.

EXAMPLE 1 Factor $x^2 + 7x + 6$.

If we want to factor $x^2 + 7x + 6$ into a product of two linear factors, then we must find numbers b and d so that $x^2 + 7x + 6 = (x + b)(x + d)$. We begin our search for b and d by observing that $b \cdot d = 6$; two possible pairs of values for b and d are therefore 3 and 2 or 6 and 1. We now test these possibilities in the expression $(x + b)(x + d)$ to see which (if any) yield $x^2 + 7x + 6$. Testing $b = 2$ and $d = 3$, we get $(x + 2)(x + 3) = x^2 + 5x + 6$; this is not the expression to be factored, so we go on. Testing $b = 1$ and $d = 6$, we get $(x + 1)(x + 6) = x^2 + 7x + 6$; the factors have been found!

In the quadratic $x^2 + 7x + 6$, all the coefficients are positive, so we needed only to consider positive numbers when factoring. Now examine what happens when one (or more) of the coefficients is negative.

EXAMPLE 2 Factor $x^2 + x - 20$.

If $x^2 + x - 20 = (x + b)(x + d)$, then $b + d = 1$ and $b \cdot d = -20$. Since $b \cdot d = -20$, there are these possibilities for b and d.

$$-1 \quad \text{and} \quad 20$$
$$1 \quad \text{and} \quad -20$$
$$-2 \quad \text{and} \quad 10$$

$$
\begin{array}{ccc}
2 & \text{and} & -10 \\
-4 & \text{and} & 5 \\
4 & \text{and} & -5
\end{array}
$$

Testing these possibilities, we see that -4 and 5 are the only values for b and d that yield $b + d = 1$; the factors are therefore $(x - 4)$ and $(x + 5)$. Thus,

$$x^2 + x - 20 = (x - 4)(x + 5)$$

Notice that, in both Examples 1 and 2, b and d are interchangeable; this happens because the coefficient of x^2 in the quadratic is 1. We turn next to another type of example. When the coefficient of x^2 is not 1, we need to use the more general equation

$$(ax + b)(cx + d) = acx^2 + (ad + bc)x + bd$$

and look for values of a, b, c, and d.

EXAMPLE 3   Factor $2x^2 - 13x + 15$.

We want $2x^2 - 13x + 15$ to equal $(ax + b)(cx + d)$. Since $a \cdot c = 2$ and 2 is prime, we must have $a = 2$ and $c = 1$ (or $a = 1$ and $c = 2$; it doesn't matter which). Then,

$$2x^2 - 13x + 15 = (2x + b)(1x + d)$$

We must now seek values for b and d that yield $b \cdot d = 15$ and $2 \cdot d + 1 \cdot b = -13$. Factoring $15 = b \cdot d$, we get the following possibilities for b and d.

$$
\begin{array}{ccc}
b = 1 & \text{and} & d = 15 \\
b = -1 & \text{and} & d = -15 \\
b = 3 & \text{and} & d = 5 \\
b = -3 & \text{and} & d = -5 \\
b = 5 & \text{and} & d = 3 \\
b = -5 & \text{and} & d = -3 \\
b = 15 & \text{and} & d = 1 \\
b = -15 & \text{and} & d = -1
\end{array}
$$

We test these possibilities in $2d + b = -13$ and find that $2(-5) + (-3) = -13$; thus, $b = -3$ and $d = -5$. Our factors are $2x - 3$ and $x - 5$. You should check the result.

$$2x^2 - 13x + 15 = (2x - 3)(x - 5)$$

There are several things to notice in this example. When there is a negative coefficient in the quadratic, we must consider negative values for b and d, even when $b \cdot d$ is positive. The sign of the middle term tells us which possibilities for b and d we need to test. In Example 3, b and d have the same sign and $2d + b$ is negative. All possibilities where b and d are both positive would make $2d + b$ positive, so they can immediately be ruled out. As you work with factoring you will develop intuition about things like this that can guide the choice of possibilities to test. Sometimes there are many choices for the coefficients of the factors; the following example illustrates how complex the situation can become.

EXAMPLE 4A   Factor $6x^2 + x - 15$.

Since $a \cdot c = 6$, the possibilities for a and c are 2 and 3 or 1 and 6; thus, we have two possible partial solutions.

$$6x^2 + x - 15 = (2x + b)(3x + d)$$

or

$$6x^2 + x - 15 = (6x + b)(x + d)$$

Now we examine $-15 = b \cdot d$ in each of these cases and, as in Example 3, there are 8 possibilities for b and d. Testing these possibilities in each of the partial solutions above, we find that the factors are $6x^2 + x - 15 = (2x - 3)(3x + 5)$.

The reader might ask how we knew to set $a = 2$ and $c = 3$ rather than $a = 3$ and $c = 2$. Since it does not matter which order the factors are written in, either assignment would work. Had we begun with the partial solution $6x^2 + x - 15 = (3x + b)(2x + d)$ and tested the various possibilities, we would have obtained $6x^2 + x - 15 = (3x + 5)(2x - 3)$. These are the same factors that we wrote in the example above, but in the opposite order.

In factoring situations where the number of choices to be tested is large, an alternate method can be more efficient. We examine this method as we use it to redo Example 4.

EXAMPLE 4B   Factor $6x^2 + x - 15$.

We want to write $6x^2 + x - 15 = (ax + b)(cx + d)$. Since $(ax + b)(cx + d) = acx^2 + (ad + bc)x + bd$, we can start by writing the middle term x in $6x^2 + x - 15$ as a sum $(ad + bc)x$. Now $adbc = acbd = 6(-15) = -90$. Thus, we want two numbers, ad and bc, so that their product is -90 and their sum is the middle coefficient 1. The numbers 10 and -9 work. Now we write

$$\begin{aligned}
6x^2 + x - 15 &= 6x^2 + (10 - 9)x - 15 && \text{(Replace 1 by } 10 - 9) \\
&= 6x^2 + 10x - 9x - 15 && \text{(Distributive property)} \\
&= 2x(3x + 5) - 3(3x + 5) && \text{(Remove common factors from first two terms and last two terms)} \\
&= (3x + 5)(2x - 3) && \text{(Remove common factor } 3x + 5)
\end{aligned}$$

The procedure above gave a factorization for the polynomial $6x^2 + x - 15$. A similar procedure can be used to factor any quadratic polynomial that is not irreducible. These are the steps in the procedure for factoring $mx^2 + nx + t$, a polynomial that is not irreducible.

1. Compute the product mt; find two numbers r and s whose product is mt and whose sum is n.
2. Replace n by $r + s$ and write $mx^2 + nx + t = mx^2 + rx + sx + t$.
3. Now remove common factors from the first two terms and from the second two terms.
4. Remove the common factor to complete factorization.

See how these four steps are applied in the following example.

EXAMPLE 5 Factor $12x^2 - 19x - 18$.

Compute $12(-18) = -216$.

Find pairs of numbers whose product is -216 and examine them to see which have a sum of -19.

$$\begin{aligned}
1(-216) &\rightarrow 1 + (-216) = -215 \\
2(-108) &\rightarrow 2 + (-108) = -106 \\
3(-72) &\rightarrow 3 + (-72) = -69 \\
4(-54) &\rightarrow 4 + (-54) = -50 \\
6(-36) &\rightarrow 6 + (-36) = -30 \\
8(-27) &\rightarrow 8 + (-27) = -19
\end{aligned}$$

Rewrite the expression and factor using the distributive property.

$$\begin{aligned}
12x^2 - 19x - 18 &= 12x^2 + (8 - 27)x - 18 \\
&= 12x^2 + 8x - 27x - 18 && \text{(Distributive property)} \\
&= 4x(3x + 2) - 9(3x + 2) \\
&= (4x - 9)(3x + 2)
\end{aligned}$$

Not every polynomial can be factored. The following example illustrates a procedure for showing that a polynomial is irreducible.

EXAMPLE 6 Factor $x^2 - 2x + 4$.

Since the coefficient of x^2 is 1, both a and c are 1 in the factorization $x^2 - 2x + 4 = (ax + b)(cx + d)$. Thus, the form of any factorization is $(x + b)(x + d)$. We now consider the possibilities for $b \cdot d = 4$. Testing 4 and 1, 2 and 2, −4 and −1, and −2 and −2, we find that none of them gives a correct factorization. Since we have exhausted all the possibilities for b and d, we conclude that the quadratic does not factor; $x^2 - 2x + 4$ is an irreducible polynomial.

USING FACTORING TO SOLVE EQUATIONS

Some equations can be solved readily by factoring. Indeed that is one reason we need skills in factoring. Here is an example. When an object is thrown straight into the air, its height at any time can be computed if we know the speed at which the object was thrown. The formula from physics is $h = st - 16t^2$ where h is height, s is the speed in feet per second with which the object was thrown, and t is time in seconds. Say a model rocket is shot at a speed of 80 feet per second. Then $h = 80t - 16t^2$. We can see, for example, that 1 second after launch ($t = 1$) the height of the rocket is $h = 80(1) - 16(1)^2 = 64$ feet; 2 seconds after launch ($t = 2$), the height is $h = 80(2) - 16(2)^2 = 96$ feet. If we want to know how long it will be before the rocket comes back to the ground, we ask, "What is t when the height h is 0?" Thus, we want to solve $0 = 80t - 16t^2$.

To solve $0 = 80t - 16t^2$, we can first factor the polynomial on the right side.

$$0 = 16t(5 - t)$$

If the product $16t(5 - t)$ is 0, then either $16t$ is 0 or $5 - t$ is 0. This observation suggests two solutions.

$$\text{If } 16t = 0 \quad \text{then } t = 0$$
$$\text{If } 5 - t = 0 \quad \text{then } t = 5$$

Thus, the rocket is on the ground at two times: $t = 0$ is the precise time at which the rocket is launched; $t = 5$ is 5 seconds after the launch.

The following exercises will help you develop your skills in factoring. Be sure to do all of them. In Chapter 10 we use factoring to solve problems similar to the one above.

Exercises 8-C

I. Factor each of the following integers into a product of two integers in as many ways as possible.

1. 12   **2.** −15

3. 63

4. −216

5. 36

6. −101

II. Factor each of the following integers into a product of prime numbers. (Use your calculator!)

1. 826 $2 \cdot 7 \cdot 59$

2. 17199

3. 4208

4. 4096

5. 2431

III. Each of the following polynomials has terms containing a common factor. Identify all common factors and factor the polynomial.

1. $4x + 8y$

2. $4x^2y + 8xy^2$

3. $-a^2b^3 + 2ab^2$

4. $-x - y$

5. $x(x + 2y) - 3y(x + 2y)$

6. $2x(x + 3) - 5(x + 3)$

7. $3x(2x - 1) + (2x - 1)$

8. $2c(a - b) + d(b - a)$

9. $(bx - by)(x^2 + 7xy + y^2)$

10. $(3x - 6)(2y + 8)$

11. $(x - y)^2b - 25(x - y)^2b^3$

12. $2xy + 2y^2 + 3x^2 + 3xy = 2y(\underline{x} + \underline{y}) + 3x(\underline{x} + \underline{y})$

13. $xy - y + x^2 - x$

14. $2y^3 + y^2 + 14y + 7$

15. $x^3 - 2x^2y + xy^2 + 2y^3$

IV. If possible, factor each of the following quadratic polynomials into a product of linear polynomials. If not possible, label the quadratic polynomial "irreducible." The first four have been started for you.

1. $x^2 + 8x + 15 = (x + \underline{5})(x + \underline{3})$

2. $x^2 - 3x - 18 = (x + \underline{3})(x - \underline{6})$

3. $3x^2 + 5x + 2 = (3x + \underline{2})(x + \underline{1})$

4. $-2x^2 - 11x + 6 = -(2x^2 + 11x - 6) = -(2x - \underline{1})(x + \underline{6})$

5. $x^2 + 6x + 5$ $(\quad)(\quad)$

6. $x^2 - 5x - 14$

7. $-x^2 - 3x + 54$

8. $x^2 + 5x - 36$

9. $2x^2 + 13x + 15$

10. $5x^2 - 10x + 4$

11. $-6x^2 - 7x + 3$

12. $4x^2 - 21x - 18$

13. $x^2 + x + 1$

14. $9x^2 - 9x - 4$

15. $9x^2 + 32x - 4$

16. $5x^2 - 44x + 32$

17. $-12x^2 - 5x + 3$

18. $14x^2 + 31x - 10$

19. $50x^2 - 25x + 3$

20. $5x^2 - 2x + 3$

21. $20x^2 - 35x + 6$

V. Realizing that $a^2 - b^2 = (a - b)(a + b)$, factor these special quadratics (often referred to as differences of squares).

1. $x^2 - 49$

2. $36 - t^2$

3. $25 - 4n^2$

4. $16 - 9y^2$

5. $r^2 - 36$

6. $a^2b^2 - 36c^2$

7. $x^4 - 81$

8. $25x^4 - y^8$

9. Can you factor the sum of two squares? For example, can you factor $x^2 + 4$ or $16 + y^2$?

NO

VI. Both the sum and difference of cubes can be factored. Verify these two equations and use them to factor the polynomials that follow.
$$a^3 - b^3 = (a - b)(a^2 + ab + b^2)$$
$$a^3 + b^3 = (a + b)(a^2 - ab + b^2)$$

1. $x^3 - 8$
2. $27 + t^3$
3. $125y^3 + 8$
4. $(a + b)^3 - c^3$

5. $40a^3 + 625b^3$
6. $1 + x^3$
7. $1 - (x + y)^3$

VII. Solve the following problems.

1. One factor of $x^4 + 2x^3 - 7x^2 - 20x - 12$ is $x^2 + 4x + 4$. Use that information to factor $x^4 + 2x^3 - 7x^2 - 20x - 12$ completely.

2. One factor of $2a^5 + a^4b - a^3b^2 - 2a^2b^3 - ab^4 + b^5$ is $a^3 - b^3$. Use that information to factor $2a^5 + a^4b - a^3b^2 - 2a^2b^3 - ab^4 + b^5$ completely.

3. If $x + 3$ is a factor of $3x^3 + 8x^2 + x + c$, what is the number c?

4. A ball is shot from a toy gun at ground level straight up into the air. If the ball is shot at a speed of 144 feet per second, how long is it before the ball comes back to the ground?

5. A stone thrown into the air hits the ground after $2\frac{1}{2}$ seconds. At what speed was the stone thrown?

VIII. Factor each of the following polynomials completely. Remember to remove common factors first.

1. $12x^2 - 10x - 12$
2. $4x^3 + 7x^2 + 3x$
3. $6x^3 + 2x^2 - 4x$
4. $r^4 - 9r^2$
5. $y^4 - 8y^2 + 16$
6. $t^4 - 10t^2 + 9$
7. $n^3 - 3n^2 + 2n$
8. $n^4 + n^3 + 5n^2$
9. $16x^2 - 25x^4$
10. $6z^2 + z - 12$

11. $x^6 - y^6$
12. $x^2y - y^3$
13. $8x^2 - 50$
14. $81m^2 - 16n^2$
15. $x^3 - 36x$
16. $4y^4 - 12y^2 + 9$
17. $8x^3 + y^3z^3$
18. $108 - 4t^3$
19. $(x + y)^3 - 1$
20. $4x^2 + 1$

REVIEW PROBLEMS

1. Find the following products.
   (a) $(4x - y)(x + 3y)$
   (b) $(3y - 2)^2$
   (c) $(x^2 + 1)(5x^2 - 2x + 1)$

2. Simplify by performing the multiplication and then combining like terms.
   (a) $(2x - y)(2x + y) + x^2 - 2y^2$
   (b) $(3x - y)^2 - (3x + y)^2$

3. One factor of $x^3 - 8x + 3$ is $x^2 - 3x + 1$. Find a second factor.

4. Factor each of the following polynomials completely.
   (a) $x^2 - 16x + 64$
   (b) $b^3 - b^2 - 20b$
   (c) $25x^2 - y^4$
   (d) $-10x^3y^2 - 2x^2y^2 + 20x^2y$
   (e) $(4a^2 + 20a + 25) - b^2$
   (f) $8y^3 + 27$
   (g) $25x^6 + y^2$
   (h) $64x^6 - y^3$

5. A book cover can be sewn from a rectangular piece of cloth. The cloth must exceed the dimensions of the book laid flat by 3 inches on each end and 1 inch on the top and the bottom. If x denotes the length of a book laid flat and y its height, write an expression for the area of a piece of cloth that can be sewed into a cover.

6. The volume of a box can be computed by finding the product of the length, width, and height of the box. A box can be made from a square piece of cardboard by first cutting a small square from each corner, turning up the sides, and then taping the edges. If the original piece of cardboard has a side length of x feet, give an expression for the volume of a box made by cutting a 1 foot square from each corner.

$25x^6 + y^2$

CHAPTER 9

ARITHMETIC OF RATIONAL EXPRESSIONS

9.1 RECOGNIZING RATIONAL EXPRESSIONS

In the last chapter we explored the arithmetic of polynomials in some detail; as we observed there, polynomials are useful in many applications of mathematics. However, not every expression we have used (or will use) in our work is a polynomial. For example, $\dfrac{1}{x}$ and $\dfrac{x^2 + 1}{x + 2}$ are not polynomials; these expressions belong to the family that we will study in this chapter, called **rational expressions.** A rational expression is a quotient of two polynomials. The following are examples of rational expressions.

$$\frac{x^3 - 3x^2 + 1}{2x + x^2}$$

$$\frac{x^5 - 1}{x}$$

$$\frac{3x^2 - 2x + 1}{1}$$

$$\frac{x^2 - 16}{x + 4}$$

$$\frac{-2}{3}$$

Notice that any rational number can be considered as a rational expression since a rational number is a quotient of integers and integers are polynomials of degree 0. A polynomial can also be considered as a rational expression; to see this, select any polynomial and divide it by 1. Although 0 is a polynomial we cannot divide by 0, so expressions like $\dfrac{x^2 + 3x + 2}{0}$ are meaningless and are not rational expressions. Throughout this chapter, and wherever rational expressions are used, you will need to remember that the denominator cannot be zero if the expression is to be meaningful.

Summarizing, we can think of a rational expression as a quotient, or a "fraction," in which the numerator and denominator are polynomials and the denominator is not the polynomial 0. Note that there are many expressions that are not rational; for example, 2^x, $\dfrac{x^2}{2^x}$, and $\sqrt{x^2 + 3}$ are not rational expressions.

Some problem situations are described by rational expressions. An example follows. In Chapter 10 you will learn to solve equations containing rational expressions in order to find solutions to certain problems.

A recreation club has "regular members" who may use all facilities and 100 "golf-only members" who may use only the golf course. The club charges a regular member a $50 annual fee. In addition, the $4000 cost of maintaining the swimming pool is divided among the regular members, and the $10,000 cost of running the golf course is divided among the regular members and the golf-only members. Write an expression for the total club charges to each of the regular members.

If x denotes the number of regular members, the cost to each regular member has these components.

Annual fee	$50.00
Pool maintenance	$\dfrac{4000}{x}$ dollars
Golf course	$\dfrac{10,000}{x + 100}$ dollars
Total costs	$50 + \dfrac{4000}{x} + \dfrac{10,000}{x + 100}$

Exercises 9-A

1. A car is traveling 20 miles per hour faster than a bus. If x denotes the speed of the bus, write an expression for the time required by the car to go 150 miles.

2. Bryon spends twice as much time on the commuter train as he does in his car on a typical work day. If x denotes the amount of time he spends in his car and if his train ride is 22 miles, write an expression for the speed of the train.

3. A boat that travels x miles per hour in still water, goes $(x + 2)$ miles per hour when it travels downstream with a 2 mile per hour current and $(x - 2)$ miles per hour when it travels upstream against a 2 mile per hour current. Write an expression that gives the time required for the boat to make a 30-mile round trip on a river with a 2 mile per hour current.

4. The Planning Committee for the spring picnic estimates that the cost of the picnic will be a $200 fixed cost, plus $0.50 per person attending. If x denotes the number of persons attending the picnic, write an expression that gives the per person cost.

5. Two sheets of paper have the same height, but the width of the second is 3 inches more than the width of the first. If w denotes the width of the first sheet and the second sheet has an area of 143 square inches, write an expression that gives the height of the two sheets.

9.2   EVALUATING RATIONAL EXPRESSIONS

The same principles that we used in evaluating the algebraic expressions in Chapter 1 are used in evaluating rational expressions. Thus, to evaluate $\frac{x^2 + 1}{x - 3}$ when $x = 2$, we replace x by 2 each time that x appears, to get $\frac{2^2 + 1}{2 - 3}$. By computing, we find that -5 is the value of $\frac{x^2 + 1}{x - 3}$ when $x = 2$.

If we try to apply this procedure to $\frac{x^2 + 1}{x - 3}$ when $x = 3$, we find that the numerator is $3^2 + 1 = 10$ and the denominator is $3 - 3 = 0$. Since $\frac{10}{0}$ has no meaning, we cannot compute a numerical value. In this case we say that $\frac{x^2 + 1}{x - 3}$ is **undefined** for $x = 3$. It sometimes happens that both the numerator and denominator of a rational expression are 0 for the same value of the variable; since $\frac{0}{0}$ is not a number, we say again that the expression is **undefined.**

Consider the expression $\frac{x^2 - 16}{x^2 - 5x + 4}$.

● If $x = -4$, then $x^2 - 16 = 0$ and $x^2 - 5x + 4 = 40$. The value of $\frac{x^2 - 16}{x^2 - 5x + 4}$ is $\frac{0}{40}$ when $x = -4$. Since 0 divided by any nonzero number is 0, the value of $\frac{x^2 - 16}{x^2 - 5x + 4}$ is 0 when $x = -4$.

● If $x = 1$, then $x^2 - 16 = -15$ and $x^2 - 5x + 4 = 0$. Thus, for $x = 1$, the expression $\frac{x^2 - 16}{x^2 - 5x + 4}$ is undefined since -15 divided by 0 is undefined.

● If $x = 4$, then $x^2 - 16 = 0$ and $x^2 - 5x + 4 = 0$, and the expression $\frac{x^2 - 16}{x^2 - 5x + 4}$ is undefined since division by 0 is undefined.

● If $x = 5$, then $x^2 - 16 = 9$ and $x^2 - 5x + 4 = 4$ and the value of $\frac{x^2 - 16}{x^2 - 5x + 4}$ is $\frac{9}{4}$.

In general, there are only a few values for which a rational expression is undefined; in fact, the number of such values can never be greater than the degree of the denominator. The following example illustrates how these values are found.

EXAMPLE Find all values of the variable for which the rational expression $\dfrac{x + 4}{2x^3 + 5x^2 - 3x}$ is undefined.

The rational expression is undefined for those values of x that make the denominator equal to 0. Thus, we must solve the equation $2x^3 + 5x^2 - 3x = 0$.

$$2x^3 + 5x^2 - 3x = 0$$

$$x(2x^2 + 5x - 3) = 0$$

$$x(2x - 1)(x + 3) = 0$$

There are three solutions to the equation.

$$x = 0 \qquad 2x - 1 = 0 \qquad x + 3 = \;\; 0$$

$$x = \frac{1}{2} \qquad x = -3$$

The rational expression is undefined when $x = 0$, $x = \dfrac{1}{2}$, or $x = -3$.

Exercises 9-B

I. Evaluate each of the following expressions for the indicated values of the variable.

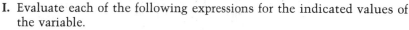

1. $\dfrac{x^2}{x + 1}$ $x = 1, 2, \dfrac{2}{3}, 0$

2. $\dfrac{y^2 - 1}{y^2 + 1}$ $y = -\dfrac{5}{2}, -1, 0, 1, \dfrac{4}{7}$

3. $\dfrac{x^2 - 5x + 4}{x - 1}$ $x = 1, 4, 10, 0$

4. $\dfrac{x^3 + x^2 + x + 1}{x^3 - x^2 - x - 1}$ $x = 0, 1, \dfrac{1}{2}, -2$

5. $\dfrac{(x + 2)^3}{x^3 + 8}$ $x = 0, -2, 2, -1$

II. Find all values of the variable for which the following rational expressions are undefined.

1. $\dfrac{x + 2}{x^2}$

2. $\dfrac{x^2 - 4x + 4}{x - 2}$

3. $\dfrac{x^2 - 9}{x(x + 3)}$

4. $\dfrac{x}{x(x^2 - 9)}$

5. $\dfrac{x^2 - x}{4}$

6. $\dfrac{15}{x + 2}$

7. $\dfrac{5x + 1}{5x + 1}$

8. $\dfrac{2x - 1}{2x^2 - 7x - 15}$

9. $\dfrac{-12}{x^2 + 11x + 24}$ 10. $\dfrac{x^2 - x}{x^3 + 3x^2 - 4x}$

III. Find all values of the variable for which the expressions in Part II are equal to zero.

IV. There should be no overlap of answers to corresponding problems in Parts II and III; explain why.

9.3   EQUIVALENCE OF RATIONAL EXPRESSIONS

Rational expressions are related to polynomials in the same way that rational numbers (or fractions) are related to integers. Computing with rational expressions is analogous to computing with fractions. (For further information about fractions refer to Appendix D.)

Recall that two fractions $\dfrac{a}{b}$ and $\dfrac{c}{d}$ are equivalent provided there are nonzero integers m and n such that $\dfrac{a \cdot m}{b \cdot m}$ and $\dfrac{c \cdot n}{d \cdot n}$ are identical. Thus, $\dfrac{9}{15}$ and $\dfrac{33}{55}$ are equivalent because $\dfrac{9 \cdot 11}{15 \cdot 11} = \dfrac{99}{165}$ and $\dfrac{33 \cdot 3}{55 \cdot 3} = \dfrac{99}{165}$. In comparing fractions, we often choose the denominators of the fractions as our numbers m and n. For example, to compare $\dfrac{5}{8}$ with $\dfrac{20}{32}$, we consider $\dfrac{5 \cdot 32}{8 \cdot 32}$ and $\dfrac{20 \cdot 8}{32 \cdot 8}$. These fractions are both equal to $\dfrac{160}{256}$; therefore, $\dfrac{5}{8} = \dfrac{20}{32}$. If we compare $\dfrac{5}{8}$ and $\dfrac{15}{32}$ using this method, we get $\dfrac{5 \cdot 32}{8 \cdot 32}$ and $\dfrac{15 \cdot 8}{32 \cdot 8}$. These fractions both have the same denominator; since their numerators are different, the fractions are not equal and $\dfrac{5}{8} \neq \dfrac{15}{32}$. Thus, there is a concise way of describing equivalent fractions: $\dfrac{a}{b} = \dfrac{c}{d}$ provided $ad = bc$. This method of showing that fractions are equivalent is discussed fully in Appendix D.

Equivalence of fractions can be extended to rational expressions; the only difference is that wherever an integer appears in the description of equivalence of fractions, it is replaced by a polynomial in the definition of equivalence of rational expressions. Consider the following examples.

EXAMPLE 1 Are $\dfrac{x+2}{x^2-4}$ and $\dfrac{x}{x^2-2x}$ equivalent?

One common denominator of these fractions is $(x^2-4)(x^2-2x)$. Using the method described above, we compare the resulting numerators by computing ad and bc. Thus,

$$ad = (x+2)(x^2-2x) = x^3 + 2x^2 - 2x^2 - 4x = x^3 - 4x$$

$$bc = (x^2-4)x = x^3 - 4x$$

Since these polynomials are identical, the rational expressions $\dfrac{x+2}{x^2-4}$ and $\dfrac{x}{x^2-2x}$ are equivalent wherever both are defined.

EXAMPLE 2 Are $\dfrac{x+1}{x^2+1}$ and $\dfrac{1}{x}$ equivalent?

In this example we compare

$$(x+1)\cdot x \quad \text{and} \quad (x^2+1)\cdot 1 \quad \text{or, in simplified form}$$
$$x^2 + x \quad \text{and} \quad x^2 + 1$$

Since these are different polynomials, the rational expressions $\dfrac{x+1}{x^2+1}$ and $\dfrac{1}{x}$ are not equivalent.

On the basis of this description of the equivalence of rational expressions, we can make some further observations about equivalence. A rational expression is equivalent to any expression found by multiplying the numerator and denominator of the original expression by the same polynomial. For example, $\dfrac{2x}{3x+2}$ is equivalent to $\dfrac{2x(x^2-5x+4)}{(3x+2)(x^2-5x+4)}$, wherever both are defined. To see that this is so, compare

$$2x(3x+2)(x^2-5x+4) \quad \text{and} \quad (3x+2)(2x)(x^2-5x+4)$$

Since both have the same factors, the products are equal and the rational expressions are equivalent. Thus, we can change a rational expression to an equivalent rational expression by multiplying numerator and denominator by the same polynomial.

There is a second observation we can make from the example $\dfrac{2x}{3x+2} = \dfrac{2x(x^2-3x+4)}{(3x+2)(x^2-3x+4)}$. The rational expression on the right has a factored numerator and a factored denominator, and the numerator and denominator share a common factor. This common factor, $x^2 - 3x + 4$, can be removed to leave the simplified equivalent expression $\dfrac{2x}{3x + 2}$.

The process of factoring numerator and denominator and removing common factors is called **reducing a rational expression.** To see how this works, consider these examples.

EXAMPLE 1   Reduce $\dfrac{x^2 - 5x + 6}{x^2 - 2x + 3}$.

$$\frac{x^2 - 5x + 6}{x^2 - 2x + 3} = \frac{(x - 3)(x - 2)}{(x - 3)(x + 1)} \qquad \text{(Factor numerator and denominator)}$$

$$= \frac{x - 2}{x + 1} \qquad \text{(Remove common factor)}$$

EXAMPLE 2   Reduce $\dfrac{x^4 - 4x^2}{x^4 + 4x^3 + 4x^2}$.

$$\frac{x^4 - 4x^2}{x^4 + 4x^3 + 4x^2} = \frac{x^2(x - 2)(x + 2)}{x^2(x + 2)^2} \qquad \text{(Factor numerator and denominator)}$$

$$= \frac{x - 2}{x + 2} \qquad \text{(Remove common factors)}$$

Exercises 9-C

I. Find three rational expressions equivalent to each of the following.

EXAMPLE: Three rational expressions equivalent to $\dfrac{x}{x + 2}$ are

$$\frac{3x}{3(x + 2)}, \qquad \frac{x^2}{x(x + 2)}, \qquad \text{and} \qquad \frac{x(x - 2)}{(x + 2)(x - 2)} = \frac{x^2 - 2x}{x^2 - 4}.$$

1. $\dfrac{x + 1}{x^2}$

2. $\dfrac{x + 2}{x - 1}$

3. $\dfrac{x^3 - 8}{x^2 + 4}$

4. $\dfrac{x^2}{2x + 2}$

5. $\dfrac{2}{5x}$

6. $3x^2$

II. Indicate whether the following pairs of expressions are equivalent.

1. $\dfrac{x^2 - 1}{x^2 + 2x + 1}$ $\qquad$ $\dfrac{x - 1}{x + 1}$

2. $\dfrac{5x + 1}{x^2}$ $\dfrac{5x^3 + x^2}{x^4}$

3. $\dfrac{x - 2}{x + 2}$ $\dfrac{x^2 - 4}{x^2 + 4x + 4}$

4. $\dfrac{3x}{x + x^2}$ $\dfrac{3x^2 + 3x}{x + 2x^2 + x^3}$

5. $\dfrac{3s^2}{s^2 + s}$ $\dfrac{3s^2 + 2}{s^2 + s + 2}$

6. $\dfrac{x^4 - 16}{x^2 + 4x + 4}$ $\dfrac{x^2 - 4}{x^2 - 4x + 4}$

7. $\dfrac{x^2 + 3x}{x^2 + 3x + 2}$ $\dfrac{1}{2}$

8. $\dfrac{8x^4}{(2x^2 + 4x)^2}$ $\dfrac{2x^2}{x^2 + 4x + 4}$

9. $\dfrac{3x + 2}{x^2}$ $\dfrac{x^2}{3x + 2}$

10. $\dfrac{x^4 - 1}{x^2 - 1}$ $x^2 + 1$

III. Reduce the following expressions by removing common factors from both numerator and denominator.

1. $\dfrac{x^7}{x^3}$ 6. $\dfrac{12x^5}{5x^{12}}$

2. $\dfrac{x^3}{x^7}$ 7. $\dfrac{24x^3}{8x}$

3. $\dfrac{y^{19}}{y^4}$ 8. $\dfrac{x^{m+2}}{x^m}$

4. $\dfrac{5y^5}{25y^2}$ 9. $\dfrac{y^n}{y^{n+5}}$

5. $\dfrac{(2x)^2}{2x^2}$ 10. $\dfrac{x^{m+n}}{x^n}$

IV. Reduce the following rational expressions by removing the common factors from numerator and denominator.

1. $\dfrac{x^3}{x^5 - x^2}$ 6. $\dfrac{x^3 - 3x^2}{3x - x^2}$

2. $\dfrac{x^2(2x + 1)^4}{x^3(2x + 1)^3}$ 7. $\dfrac{x^2 - 5x - 6}{x^2 - 4x - 5}$

3. $\dfrac{x^2 - 4}{x^2 + 4x + 4}$ 8. $\dfrac{x^4 - 16}{x^4 - 5x^2 + 4}$

4. $\dfrac{x + 1}{x^2 + 1}$ 9. $\dfrac{x^2(x - 1)(x + 2)}{x^2(x + 1)(x - 2)}$

5. $\dfrac{x^5 - 9x^3}{3x^2}$ 10. $\dfrac{(x + 1)^2(x - 4)^2(x + 3)^3}{(x + 1)(x - 1)(x^2 - 16)(x + 3)^5}$

9.4 MULTIPLICATION AND DIVISION OF RATIONAL EXPRESSIONS

The methods of finding products, quotients, sums, and differences of rational expressions are very similar to the analogous operations for fractions (described in Appendix D). We recall with the following examples how these operations are defined for fractions.

$$\frac{2}{3} \cdot \frac{4}{5} = \frac{2 \cdot 4}{3 \cdot 5} = \frac{8}{15}$$

$$\frac{2}{3} \div \frac{4}{5} = \frac{2}{3} \cdot \frac{5}{4} = \frac{2 \cdot 5}{3 \cdot 4} = \frac{10}{12}$$

$$\frac{2}{3} + \frac{4}{5} = \frac{2 \cdot 5}{3 \cdot 5} + \frac{4 \cdot 3}{5 \cdot 3} = \frac{10 + 12}{15} = \frac{22}{15}$$

$$\frac{2}{3} - \frac{4}{5} = \frac{2 \cdot 5}{3 \cdot 5} - \frac{4 \cdot 3}{5 \cdot 3} = \frac{10 - 12}{15} = \frac{-2}{15}$$

The methods of computation for rational expressions combine three elements: (1) the arithmetic of fractions, (2) the arithmetic of polynomials that we covered in Chapter 8, and (3) attention to the values of variables for which the expressions are not defined.

Suppose we wish to multiply $\frac{5x + 2}{x - 1}$ by $\frac{2x^2}{3x - 4}$. Applying the rules for multiplication of fractions and multiplication of polynomials, we obtain the following product.

$$\frac{5x + 2}{(x - 1)} \cdot \frac{2x^2}{3x - 4} = \frac{(5x + 2) \cdot 2x^2}{(x - 1)(3x - 4)} \qquad \text{(Multiplication of fractions)}$$

$$= \frac{10x^3 + 4x^2}{3x^2 - 7x + 4} \qquad \text{(Multiplication of polynomials)}$$

This product is defined except when $x - 1 = 0$ and when $3x - 4 = 0$, that is, whenever both of the original fractions are defined. Thus, the product is defined whenever $x \neq 1$ and $x \neq \frac{4}{3}$.

Here is another example.

$$\frac{x^2 - 4}{2x + 5} \cdot \frac{2x - 5}{x - 2} = \frac{(x^2 - 4)(2x - 5)}{(2x + 5)(x - 2)} \quad \text{when } x \neq -\frac{5}{2} \text{ and } x \neq 2$$

There are two things that you should notice. The first is that the product of two rational expressions is a rational expression. The second is that this way of multiplying rational expressions "agrees with" everything else we have done so far. What we mean by agreement is illustrated by the following example. Suppose you were given the task "find the product of $\dfrac{x-2}{5x}$ and $\dfrac{3x+2}{x-1}$ when $x = 3$." There are two ways you might attack this job.

METHOD 1 Multiply the rational expressions and evaluate the product at $x = 3$.

Multiply:
$$\frac{x-2}{5x} \cdot \frac{3x+2}{x-1} = \frac{(x-2)(3x+2)}{5x(x-1)}$$

$$= \frac{3x^2 - 4x - 4}{5x^2 - 5x}$$

Evaluate:
$$\frac{3(3)^2 - 4(3) - 4}{5 \cdot 3^2 - 5 \cdot 3} = \frac{27 - 12 - 4}{45 - 15} = \frac{11}{30}$$

METHOD 2 Evaluate each rational expression at $x = 3$, then multiply the results.

Evaluate:
$$\frac{3-2}{5 \cdot 3} = \frac{1}{15}$$

$$\frac{3 \cdot 3 + 2}{3 - 1} = \frac{11}{2}$$

Multiply:
$$\frac{1}{15} \cdot \frac{11}{2} = \frac{11}{30}$$

Both methods yield the same result, not only in this case, but also wherever all of the rational expressions involved are defined. A similar result will hold for all the operations on rational expressions. You will see in the problems at the end of this chapter that if you add, subtract, multiply, or divide two rational expressions and evaluate the result at some value of the variable you will get the same result that arises from first evaluating each expression at that value and then performing the operation.

As with multiplication, the division of rational expressions is also performed like that of fractions. We consider a few examples.

EXAMPLE 1
$$\frac{x}{x-4} \div \frac{x-2}{x+5} = \frac{x}{x-4} \cdot \frac{x+5}{x-2} \qquad \text{(Division of fractions)}$$

$$= \frac{x(x+5)}{(x-4)(x-2)} \qquad \text{(Multiplication of fractions)}$$

This division has meaning except at those values of x where one of the rational expressions in the computation is undefined; that is, except where

$$x - 4 = 0 \quad \text{or} \quad x = \quad 4 \qquad \text{(First expression and quotient}$$
$$\text{undefined)}$$
$$x + 5 = 0 \quad \text{or} \quad x = -5 \qquad \text{(Second expression undefined)}$$
$$x - 2 = 0 \quad \text{or} \quad x = \quad 2 \qquad \text{(Quotient undefined)}$$

EXAMPLE 2 $\dfrac{x^3}{2x - 3} \div \dfrac{2x - 3}{x^3} = \dfrac{x^3}{2x - 3} \cdot \dfrac{x^3}{2x - 3}$

$$= \dfrac{x^6}{(2x - 3)^2}$$

This division has meaning except when

$$2x - 3 = 0 \quad \text{or} \quad x = \frac{3}{2} \qquad \text{(First expression and quotient}$$
$$\text{undefined)}$$
$$x^3 = 0 \qquad \text{or} \quad x = 0 \qquad \text{(Second expression undefined)}$$

EXAMPLE 3 $\dfrac{x^2 - 5x}{2x + 1} \div \dfrac{x^2 + 2x}{3x - 1} = \dfrac{(x^2 - 5x)(3x - 1)}{(2x + 1)(x^2 + 2x)}$

$$= \dfrac{3x^3 - 16x^2 + 5x}{2x^3 + 5x^2 + 2x}$$

This division has meaning except when

$$2x + 1 = 0 \quad \text{or} \quad x = -\frac{1}{2} \qquad \text{(First expressoin and quotient}$$
$$\text{undefined)}$$

$$3x - 1 = 0 \quad \text{or} \quad x = \frac{1}{3} \qquad \text{(Second expression undefined)}$$

$$x^2 + 2x = 0 \quad \text{or} \quad x = 0, x = -2 \quad \text{(Quotient undefined)}$$

Examining these three examples, we make the following observations. The quotient of two rational expressions is a rational expression; the quotient is defined wherever the expressions that are being divided are defined and the divisor is not zero.

Exercises 9-D

I. Find the products and quotients indicated; give the values of the variable that are excluded.

1. $\dfrac{x + 2}{x + 3} \cdot \dfrac{x}{x + 1}$

3. $\dfrac{3x + 2}{5x + 4} \cdot 7x^2$

2. $\dfrac{x^2 + 1}{x^2} \cdot \dfrac{x^3}{x^2 - 1}$

4. $\dfrac{2x + 3}{x^3} \cdot \dfrac{x^3}{2x - 3}$

5. $\dfrac{1}{x^3 - 1} \cdot \dfrac{x^3}{3x^2 - 1}$

6. $\dfrac{x^2 - 9}{x^2 + 9} \cdot \dfrac{x + 3}{x - 3}$

7. $\dfrac{x}{x + 1} \cdot \dfrac{x + 2}{x + 3} \cdot \dfrac{x}{x + 4}$

8. $\dfrac{x^3}{x^2 - 1} \cdot \dfrac{x^3 - 1}{x^2}$

9. $3x^2 \cdot \dfrac{1}{5x^4}$

10. $\dfrac{x^2 + 7x + 6}{x^2 + 3x + 2} \cdot \dfrac{x^2 + x - 2}{x^2 + 5x - 6}$

11. $\dfrac{x^2}{x + 1} \div \dfrac{3x^2}{x - 1}$

12. $\dfrac{x^4}{x^2 + 3} \div \dfrac{x + 3}{x^4}$

13. $\dfrac{2x + 3}{x^3 - 1} \div \dfrac{2x + 3}{x^3 - 1}$

14. $\dfrac{x^2 + 3x + 2}{x^2 + 3x - 4} \div \dfrac{x + 2}{x - 4}$

15. $\dfrac{x^2 - 3x - 4}{x^7} \div \dfrac{x^6}{x^2 - 3x - 4}$

16. $\dfrac{x^2}{x + 1} \cdot \dfrac{3x}{2x + 1} \div \dfrac{x^3}{x + 1}$

17. $\dfrac{5x}{3x + 1} \div \dfrac{x^2}{x + 3} \div \dfrac{1}{x + 3}$

18. $\dfrac{x^2}{x^2 + 1} \div \dfrac{x^2 + 1}{x^2} \cdot \dfrac{x^2}{x^2 + 1}$

19. $\dfrac{(x^2 + 1)(x - 2)}{x^3(x - 5)} \div \dfrac{(x^2 + 1)x^3}{(x - 2)(x - 5)}$

20. $\dfrac{(x^2 + 1)(x - 2)}{x^3(x - 5)} \cdot \dfrac{(x^2 + 1)x^3}{(x - 2)(x - 5)}$

II. For each of the following, evaluate in two ways and show that the results agree.

1) substitute
2) simplify, then substitute

1. $\dfrac{x}{x + 3} \cdot \dfrac{x^2 - 9}{x^2}$ at $x = 3$

2. $\dfrac{x}{x + 3} \cdot \dfrac{x^2 - 9}{x^2}$ at $x = -3$

3. $\dfrac{x^2 + 4}{x - 3} \cdot \dfrac{x^2 - 5x + 6}{x + 4}$ at $x = 2$

4. $\dfrac{x^2}{x + 1} \div \dfrac{x - 1}{x^2}$ at $x = -1$

5. $\dfrac{x + 1}{x - 2} \div \dfrac{x + 3}{x - 4}$ at $x = -1, 2, -3,$ and 4

9.5 ADDITION AND SUBTRACTION OF RATIONAL EXPRESSIONS

The procedures for adding and subtracting rational expressions, like those for multiplying and dividing, are almost identical to the rules for fractions (see Appendix D for details). Let us recall the steps in the addition of two fractions, $\dfrac{3}{4}$ and $\dfrac{2}{5}$.

$\dfrac{3}{4}$ is equivalent to $\dfrac{3 \cdot 5}{4 \cdot 5} = \dfrac{15}{20}$.

$\dfrac{2}{5}$ is equivalent to $\dfrac{2 \cdot 4}{5 \cdot 4} = \dfrac{8}{20}$.

20 is a common denominator.

$$\frac{3}{4} + \frac{2}{5} = \frac{15}{20} + \frac{8}{20} = \frac{23}{20}.$$

The essential step in these procedures is to find fractions with the same denominator that are equivalent to the ones to be added. Then we can add the numerators of the fractions found. We follow these steps in the following examples.

EXAMPLE 1   Find the sum $\dfrac{x^2}{x + 1} + \dfrac{2x}{x + 1}$.

These expressions have a common denominator, so

$$\frac{x^2}{x + 1} + \frac{2x}{x + 1} = \frac{x^2 + 2x}{x + 1}$$

EXAMPLE 2   Find the sum $\dfrac{x^2}{2} + \dfrac{2}{3x}$.

Choose $2(3x)$ as a common denominator.

$$\frac{x^2}{2} \text{ is equivalent to } \frac{x^2 \cdot 3x}{2 \cdot 3x}$$

$$\frac{2}{3x} \text{ is equivalent to } \frac{2 \cdot 2}{2 \cdot 3x}$$

Therefore,

$$\frac{x^2}{2} + \frac{2}{3x} = \frac{x^2 \cdot 3x}{2 \cdot 3x} + \frac{2 \cdot 2}{2 \cdot 3x} = \frac{3x^3 + 4}{6x} \qquad \text{for } x \neq 0$$

EXAMPLE 3   Find the sum $\dfrac{x^3}{x^2 + 1} + 5x$.

Consider $5x$ as $\dfrac{5x}{1}$ and choose $1(x^2 + 1)$ as a common denominator.

$$\frac{5x}{1} \text{ is equivalent to } \frac{5x(x^2 + 1)}{x^2 + 1}$$

Therefore,

$$\frac{x^3}{x^2 + 1} + 5x = \frac{x^3}{x^2 + 1} + \frac{5x^3 + 5x}{x^2 + 1} = \frac{6x^3 + 5x}{x^2 + 1}$$

Recall that $x^2 + 1$ is never 0, so this addition makes sense for all values of x.

In Examples 2 and 3 we chose the common denominator to be the product of the denominators in the rational expressions being added. That is one way to produce a common denominator. If the denominators in the rational expressions being added have several factors, it may be more efficient to factor the denominators and build a common denominator from the factors. The following subtraction example illustrates that idea. Subtraction of rational expressions is performed in the same way as addition.

EXAMPLE 4   Find the difference $\dfrac{x}{x^2 - 4} - \dfrac{x}{x^2 - x - 6}$.

The denominators can be factored in this way.

$$x^2 - 4 = (x - 2)(x + 2)$$

$$x^2 - x - 6 = (x - 3)(x + 2)$$

The polynomial $(x - 2)(x + 2)(x - 3)$ is a multiple of both $x^2 - 4$ and $x^2 - x - 6$. Thus, it can be used as a common denominator.

$$\frac{x}{x^2 - 4} = \frac{x}{(x - 2)(x + 2)} = \frac{x(x - 3)}{(x - 2)(x + 2)(x - 3)}$$

$$\frac{x}{x^2 - x - 6} = \frac{x}{(x - 3)(x + 2)} = \frac{x(x - 2)}{(x - 2)(x + 2)(x - 3)}$$

Therefore,

$$\frac{x}{x^2 - 4} - \frac{x}{x^2 - x - 6} = \frac{x}{(x - 2)(x + 2)} - \frac{x}{(x - 3)(x + 2)}$$

$$= \frac{x(x - 3)}{(x - 2)(x + 2)(x - 3)}$$

$$- \frac{x(x - 2)}{(x - 2)(x - 3)(x + 2)}$$

$$= \frac{x(x - 3) - x(x - 2)}{(x - 2)(x + 2)(x - 3)}$$

$$= \frac{x^2 - 3x - x^2 + 2x}{(x - 2)(x + 2)(x - 3)}$$

$$= \frac{-x}{(x - 2)(x + 2)(x - 3)}$$

COMPLEX FRACTIONS

You will sometimes see fractional expressions that have sums and differences of rational expressions in their numerators and denominators. Two common ways of simplifying these **complex fractions** are illustrated in the following example. We want to rewrite the fraction so that the numerator and denominator are polynomials.

EXAMPLE 5   Simplify $\dfrac{\dfrac{1}{x} - \dfrac{1}{x^2}}{x - \dfrac{1}{x}}$.

METHOD 1   We can write the numerator as a single fraction and the denominator as a single fraction, and then divide the numerator by the denominator.

Numerator: $\dfrac{1}{x} - \dfrac{1}{x^2} = \dfrac{x}{x^2} - \dfrac{1}{x^2} = \dfrac{x - 1}{x^2}$

Denominator: $x - \dfrac{1}{x} = \dfrac{x^2}{x} - \dfrac{1}{x} = \dfrac{x^2 - 1}{x}$

$$\frac{\dfrac{1}{x} - \dfrac{1}{x^2}}{x - \dfrac{1}{x}} = \frac{\dfrac{x - 1}{x^2}}{\dfrac{x^2 - 1}{x}}$$

$$= \frac{x - 1}{x^2} \div \frac{x^2 - 1}{x}$$

$$= \frac{x - 1}{x^2} \cdot \frac{x}{x^2 - 1}$$

$$= \frac{x(x - 1)}{x^2(x - 1)(x + 1)}$$

$$= \frac{1}{x(x + 1)}$$

METHOD 2   We can multiply the numerator and denominator of the complex function by the same expression specially chosen to convert the numerator and the denominator to polynomials. In this example, multiplying by x^2 has this effect.

$$\frac{\dfrac{1}{x} - \dfrac{1}{x^2}}{x - \dfrac{1}{x}} = \frac{x^2 \left(\dfrac{1}{x} - \dfrac{1}{x^2} \right)}{x^2 \left(x - \dfrac{1}{x} \right)}$$

$$= \frac{\dfrac{x^2}{x} - \dfrac{x^2}{x^2}}{x^3 - \dfrac{x^2}{x}}$$

$$= \frac{x - 1}{x^3 - x}$$

$$= \frac{x - 1}{x(x - 1)(x + 1)}$$

$$= \frac{1}{x(x + 1)}$$

Exercises 9-E

I. Find the sums and differences indicated. Give the values of the variable that are excluded.

1. $\dfrac{x + 1}{x^3} + \dfrac{2x + 3}{x^3}$

2. $\dfrac{5x^2}{3x^2 - x + 1} + \dfrac{2x + 1}{3x^2 - x + 1}$

3. $\dfrac{1}{x} + \dfrac{x + 1}{x^2}$

4. $\dfrac{3x + 2}{x + 1} + \dfrac{2x + 3}{x - 1}$

5. $\dfrac{x^2}{2x + 1} + \dfrac{2x + 1}{x^2}$

6. $x + \dfrac{1}{x}$

7. $\dfrac{1}{x} + \dfrac{1}{x + 1} + \dfrac{1}{x - 1}$

8. $\dfrac{2x + 3}{3x + 2} + \dfrac{3x + 2}{2x + 3}$

9. $x + \dfrac{1}{x} + \dfrac{1}{x - 1}$

10. $\dfrac{x + 1}{5x^2} - \dfrac{2x}{5x^2}$

11. $\dfrac{x^2}{2x + 3} - \dfrac{3x}{2x + 3}$

12. $\dfrac{x^3}{x - 3} - \dfrac{x - 3}{x^3}$

13. $\dfrac{2x + 1}{x - 2} - \dfrac{x + 2}{2x - 1}$

14. $x - \dfrac{1}{x - 1} + \dfrac{2}{x^2 - 1}$

15. $\dfrac{3x^2}{2x + 1} - \dfrac{3x^2}{2x - 1} + \dfrac{9x^2}{4x^2 - 1}$

II. Simplify the following expressions by performing the operations indicated.

1. $\dfrac{x}{x^2 - 1} \cdot \left(\dfrac{2x - 1}{x + 1} + \dfrac{3x}{x - 1} \right)$

2. $\dfrac{3x}{x + 2} - \dfrac{2x}{x^2 - 4} \div \dfrac{x + 2}{x - 2}$

3. $\dfrac{x}{x + 1} + \dfrac{x + 1}{x + 2} \cdot \dfrac{x + 2}{x + 3} - \dfrac{x}{x - 4}$

4. $\left(\dfrac{1}{x^2 - 4} - \dfrac{4}{x^2 - 5x + 6} \right) \cdot \dfrac{x - 2}{x}$

5. $x - \dfrac{x}{2 - x} \cdot \dfrac{x - 2}{x^2}$

6. $\dfrac{\dfrac{1}{x^2} - 4}{\dfrac{1}{x} + 2}$

7. $\dfrac{1 + \dfrac{1}{x^2}}{1 + \dfrac{1}{x}}$

8. $\dfrac{\dfrac{1}{x} + \dfrac{1}{y}}{\dfrac{1}{x} - \dfrac{1}{y}}$

III. Perform the indicated operations on the rational expressions and evaluate the result at the value given.

1. $\dfrac{x}{x + 2} + \dfrac{x}{x^2 - 4}$ at $x = 1$

2. $\dfrac{x + 1}{x^2} - \dfrac{x + 2}{x}$ at $x = 3$

3. $\dfrac{x^2}{(x - 2)(x - 3)} + \dfrac{x}{(x + 2)(x - 3)}$ at $x = 1$

4. $\dfrac{x}{x + 2} + \dfrac{x + 3}{x + 4}$ at $x = -1$

5. $\dfrac{1}{x} + \dfrac{1}{x^3} + \dfrac{1}{x + x^3}$ at $x = 1$

6. $\dfrac{1}{x - 2} + \dfrac{x - 2}{x - 4}$ at $x = 2$

IV. Return to Part III; this time evaluate each rational expression at the value given, and then perform the indicated operations. Compare your results with those of Part III.

V. Evaluate each of the following at the value given.

1. $\dfrac{x}{2} \cdot \left(\dfrac{2 + x}{x} + \dfrac{x^2}{x - 1} \right)$ at $x = 2$

2. $\dfrac{x^2 - 4}{x^2 + 3x + 2} \cdot \left(\dfrac{x - 1}{x - 2} + \dfrac{x + 1}{x + 2} \right)$ at $x = 3$

3. $\dfrac{x^2 - 3x + 1}{x^5 + 4x^3} \cdot \left(\dfrac{x^2 - 2x + 1}{x + 8} - \dfrac{19x + 2}{23x + 7} \right)$ at $x = 0$

REVIEW PROBLEMS

1. Reduce the following fractions.

(a) $\dfrac{6x^5}{2x^9}$

(b) $\dfrac{x^2 - 4x + 3}{x^2 + 3x - 4}$

2. Evaluate each of the following at the values given.

(a) $\dfrac{x^2 + 2x + 1}{x^2 + 1}$ at $x = -2, -1, 0, 1,$ and 4

(b) $\dfrac{y^2 + 4y - 5}{2y^2 - 50}$ at $y = -10, -5, 0, 5,$ and 10

(c) $\dfrac{(x + 2)^3}{x^3 + 8}$ at $x = 2, 0, -2,$ and 4

3. Find the products and quotients. Leave answers in reduced form.

(a) $\dfrac{x + 3}{x + 1} \cdot \dfrac{x - 1}{x - 3}$

(b) $\dfrac{x^2 - 4}{x^2 - 1} \cdot \dfrac{x^2 - 3x + 2}{x^2 + 3x + 2}$

(c) $\dfrac{x^3}{x^2-1} \div \dfrac{x^2+1}{x^5}$

(d) $\dfrac{x^3-x}{3x^4} \div \dfrac{x^2-2x+1}{2x^3}$

4. Find the sums and differences. Leave answers in reduced form.

can't factor sum of 2 squares

(a) $\dfrac{x}{2} + \dfrac{1}{2x}$ $\dfrac{x^2}{2x} + \dfrac{1}{2x} = \dfrac{x^2+1}{2x}$

(c) $\dfrac{x+1}{x+2} - \dfrac{x+3}{x+4}$

(b) $\dfrac{3x}{x^2-1} + \dfrac{x+2}{x^2-2x+1}$

5. Evaluate the following expression at $x = 3$.

$$\dfrac{x}{4} \cdot \dfrac{x-1}{x} + \dfrac{x+2}{x^2}$$

6. Simplify the following expressions.

(a) $\dfrac{\dfrac{x^2}{x+2}}{\dfrac{x^3}{x-2}}$

*wrong answer in key

I corrected it.*

(b) $\dfrac{\dfrac{1}{x^2} - \dfrac{1}{(x-1)^2}}{\dfrac{2x-1}{x^2-x}}$

undefined

_____ *zero*

CHAPTER 10

SQUARE ROOT AND QUADRATIC EQUATIONS

10.1   PYTHAGOREAN THEOREM

There is a property of right triangles that was known to the ancient Egyptians and Babylonians and that has many mathematical applications. We are particularly interested in its application for measuring distance. Because a proof is credited to Pythagoras and his school in the fifth century B.C., the principle is commonly called **the Pythagorean theorem.** To understand what it says, you need to realize that in a right triangle the side opposite the right angle is called the **hypotenuse** and the other sides are called **legs.**

Figure 10.1

PYTHAGOREAN THEOREM

If a right triangle has legs of lengths a and b and a hypotenuse of length c, then

$$c^2 = a^2 + b^2$$

Look at Figure 10.1.

Actually, the Pythagorean theorem is a statement about areas of squares. It says that the area of a square, one side of which is the hypotenuse of a right triangle, equals the sum of the areas of the two squares that have as sides the legs of the triangle. Examine Figure 10.2. Therefore, if we know two of the three areas, we can find the third using the relationship $c^2 = a^2 + b^2$.

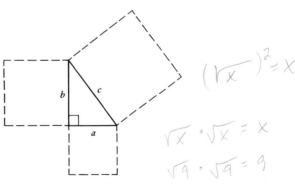

Figure 10.2

Often our interest is more in the length of the side of a right triangle than it is in the area of a square along one of the sides. This is particularly true if we need to compute distance. Here are two examples.

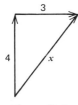

Figure 10.3

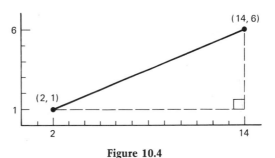

A bicyclist travels 4 miles north and 3 miles east. How far is she from her starting point?

The question here actually is how far "as the crow flies." The bicycle trip can be represented by arrows as shown in Figure 10.3, with the vertical arrow denoting the direction north and the horizontal arrow the direction east. We want to determine the distance x. Since going north and then east determines a right angle, the Pythagorean theorem says that $x^2 = 4^2 + 3^2$ or $x^2 = 25$. Because $x^2 = 25$, we conclude that the distance x is 5 miles.

Consider the two points (2, 1) and (14, 6). What is the distance between them?

We need to find the length of the segment joining the two points. We regard it as the hypotenuse of a right triangle as shown in Figure 10.4. The horizontal leg has a length of $14 - 2$ (the difference of the first coordinates) and the vertical leg has a length of $6 - 1$ (the difference of the second coordinates). Thus, the hypotenuse must have length c where $c^2 = 12^2 + 5^2 = 144 + 25 = 169$. We can conclude that $c = 13$.

Figure 10.4

In both of the examples above we were able to determine a distance because we had a value for its square. Before we consider other problems that use the Pythagorean theorem, we need to look carefully at the issue of finding a number when its square is known.

10.2 SQUARE ROOT

A concrete question that can be asked is, "If you know the area of a square, can you find the length of its side?" If a square has an area of 25, for example, we know it has a side length of 5. If a square has an area of 16, it must have a side length of 4.

When we ask these questions, we are actually asking for solutions to the equations $x^2 = 25$ and $x^2 = 16$. If we want all possible solutions (and not just the length of a side of a square), we would say that $x^2 = 25$ has the two solutions $x = 5$ and $x = -5$. The equation $x^2 = 16$ has two solutions, $x = 4$ and $x = -4$. The numbers 5 and -5 are said to be square roots of 25; 4 and -4 are square roots of 16. In general, a solution to the equation $x^2 = n$ for some number n is called a **square root** of n. Notice that if n is a negative number there is no real number for its square root.

Although a positive number has both a positive square root and a negative square root, we frequently want to designate the positive one, and we have a special symbol for it: $\sqrt{n}$ means the positive number whose square is n. Thus, $\sqrt{100} = 10$ and $\sqrt{25} = 5$. We also write $\sqrt{0} = 0$, but there is no real number $\sqrt{-25}$ because the square of any positive or negative number is positive. The symbol $\sqrt{n}$ is read as the "square root n," or "radical n," or sometimes "root n."

If n is a positive number, then both $\sqrt{n}$ and $-\sqrt{n}$ are solutions to $x^2 = n$. Is it possible that there are other solutions? The answer is no, and the argument is not difficult. If a and b are both solutions, then $a^2 = n$ and $b^2 = n$, so $a^2 = b^2$. Thus, $a^2 - b^2 = 0$ so $(a - b)(a + b) = 0$. This tells us that either $a - b = 0$ or $a + b = 0$; so either $a = b$ or $a = -b$. The argument demonstrates that if two numbers are both solutions to $x^2 = n$, then either they are equal or one is the negative of the other. Thus, $\sqrt{n}$ and $-\sqrt{n}$ are the only solutions to $x^2 = n$.

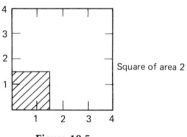

Square of area 2

Figure 10.5

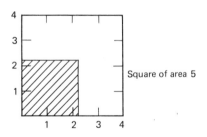

Square of area 5

Figure 10.6

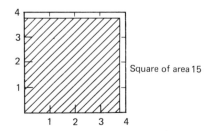

Square of area 15

Figure 10.7

We know there are numbers $\sqrt{2}$, $\sqrt{5}$, and $\sqrt{15}$ because they give us the lengths of sides of squares that have areas of 2, 5, and 15 as shown in Figures 10.5, 10.6, and 10.7. But how do we find the values of these numbers? Often we simply leave the symbols $\sqrt{2}$, $\sqrt{5}$, and $\sqrt{15}$ to denote these numbers, or we can seek numerical approximations. For example, if we want to estimate the value of $\sqrt{2}$, we might first observe that it is between 1 and 2.

Number	Number Squared
1	1
2	4

Our next try might be 1.5. The square of 1.5 is 2.25; too big. Try 1.4.

Number	Number Squared
1	1
1.4	1.96
1.5	2.25
2	4

The information above demonstrates that the number we seek is between 1.4 and 1.5, and it appears that the number is closer to 1.4. Next we might use a calculator to compute the square of 1.41. It is too small. Try 1.42.

Number	Number Squared
1	1
1.4	1.96
1.41	1.9881
1.42	2.0164
1.5	2.25
2	4

Now we see that $\sqrt{2}$ is between 1.41 and 1.42. Both are good approximations to $\sqrt{2}$, but neither equals $\sqrt{2}$. In fact, you should see that any finite decimal we would write starting with 1.41 . . . could not square to give the whole number 2, because you could write down the product and see nonzero digits in its decimal part. If we want to designate the square root of 2 exactly, we write $\sqrt{2}$; when we write a decimal we are giving an approximation. Before calculators were widely available, students used extensive tables to look up approximations to square roots of numbers or they used a complicated algorithm to compute them. Now if you have a calculator with a $\boxed{\sqrt{}}$ key that information is available to you at the push of a button.

There are two observations about square roots that will assist in computation. Because $\sqrt{n}$ is a solution to $x^2 = n$, we know $(\sqrt{n})^2 = n$. Here is our first rule.

RULE 1 FOR RADICALS   (Squaring a radical)

If n is a positive number or 0, then $(\sqrt{n})^2 = n$.

From Rule 1 we can say

$$(\sqrt{101})^2 = 101 \qquad (\sqrt{14xy})^2 = 14xy \qquad (\sqrt{a^2 + b^2})^2 = a^2 + b^2$$

The first rule tells about $\sqrt{n} \cdot \sqrt{n}$; the second rule concerns $\sqrt{a} \cdot \sqrt{b}$ for any positive numbers a and b. One solution to the equation $x^2 = ab$ is $\sqrt{ab}$. Consider also $\sqrt{a} \cdot \sqrt{b}$. Since $(\sqrt{a} \cdot \sqrt{b})^2 = (\sqrt{a})^2 \cdot (\sqrt{b})^2 = a \cdot b$, we see that $\sqrt{a} \cdot \sqrt{b}$ is also a solution to $x^2 = ab$. Since neither solution is negative, it must be that $\sqrt{a} \cdot \sqrt{b} = \sqrt{ab}$. This is the second rule.

RULE 2 FOR RADICALS   (Multiplying radicals)

If a and b are positive numbers or 0, then

$$\sqrt{a} \cdot \sqrt{b} = \sqrt{ab}$$

The second rule read from right to left permits us to write the square roots of large numbers in terms of the square roots of their factors and in this way to simplify. For example,

$$\sqrt{48} = \sqrt{16} \cdot \sqrt{3} = \sqrt{4^2} \cdot \sqrt{3} = 4 \cdot \sqrt{3}$$
$$\sqrt{45} = \sqrt{9} \cdot \sqrt{5} = \sqrt{3^2} \cdot \sqrt{5} = 3\sqrt{5}$$
$$\sqrt{4a^2b^5} = \sqrt{4} \cdot \sqrt{a^2} \cdot \sqrt{b^4} \cdot \sqrt{b} = 2ab^2 \cdot \sqrt{b}$$
$$\sqrt{27a^3s^8} = \sqrt{27} \cdot \sqrt{a^3} \cdot \sqrt{s^8} = \sqrt{9} \cdot \sqrt{3} \cdot \sqrt{a} \cdot \sqrt{a^2} \cdot \sqrt{s^8}$$
$$= 3\sqrt{3} \cdot \sqrt{a} \cdot a \cdot s^4$$
$$= 3\sqrt{3a} \cdot as^4$$

The examples show that simplifying a square root involves (1) identifying factors that occur an even number of times and (2) taking the square root of those factors.

Exercises 10-A

I. Find approximate solutions to the following equations by using your calculator, but not by using the $\sqrt{}$ key.

1. $x^2 = 169$ 4. $x^2 = -4$

2. $x^2 = 3$ 5. $x^2 = 12$

3. $4x^2 = 12$ 6. $x^2 = 5$

II. Simplify, assuming all variables are positive.

1. $\sqrt{a^8 b^4}$ 6. $(\sqrt{32})^3$

2. $\sqrt{a^{2n}}$ 7. $(\sqrt{6} - \sqrt{2})(\sqrt{6} + \sqrt{2})$

3. $\sqrt{a^3}$ 8. $\sqrt{18a^3} - a\sqrt{8a}$

4. $\sqrt{\dfrac{x^2 y^3}{z^5}}$ 9. $\sqrt{3ab} - \sqrt{27ab^3}$

5. $\sqrt{81t^{12}s^3}$ 10. $(\sqrt{5 \cdot 36 \cdot 63})^2$

     11. $\sqrt{20} + \sqrt{5} + \sqrt{45}$

III. Evaluate the expression $\dfrac{-b + \sqrt{b^2 - 4ac}}{2a}$ for the following values of a, b, and c. (This expression arises in the next section. We need to know how to evaluate it.)

1. $a = 2$, $b = 3$, $c = -2$ 3. $a = 1$, $b = -2$, $c = -6$

2. $a = 3$, $b = -1$, $c = -2$ 4. $a = 2$, $b = -3$, $c = 4$

IV. Sketch the graph of $y = \sqrt{x}$ for $x \geq 0$. Include several points for x between 0 and 1. Also graph $y = x^2$ and compare the two graphs.

V. A right triangle has legs of length a and b and the hypotenuse of length c. Find the values of a, b, and c in the following cases.

1. $a = 12$, $b = 35$, $c = \,$?   3. $a = \sqrt{10}$, $b = \,$?, $c = \sqrt{30}$

2. $a = \,$?, $b = 5$, $c = 15$ 4. $a = \,$?, $b = \sqrt{\dfrac{3}{4}}$, $c = \sqrt{\dfrac{9}{4}}$

VI. Solve the following problems.

1. At 12 noon a boat heading north at a rate of 5 miles per hour passes a buoy. At 1:00 P.M. another boat passes the buoy heading west at a rate of 4 miles per hour. What is the distance between the boats at 3:00 P.M.?

2. In a certain right triangle the hypotenuse is 16 feet longer than the shortest side. The third side is 24 feet. Give the lengths of the three sides of the triangle.

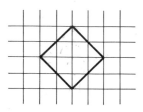

Figure 10.8

VII. Use the Pythagorean theorem to solve the following problems.

1. Find the distance between the points $(3, -3)$ and $(1, 5)$, $(10, -2)$ and $(-1, 1)$, $(6.4, 3.9)$ and $(-1.1, 4.3)$.

2. In problem 1 you computed the distance between pairs of points. The examples illustrate that the square of the distance between two points equals the square of the difference of the first coordinates plus the square of the difference of the second coordinates. Draw a picture that shows that if d is the distance between (r, s) and (u, v), then

$$d^2 = (r - u)^2 + (s - v)^2 \quad \text{and} \quad d = \sqrt{(r - u)^2 + (s - v)^2}$$

VIII. Solve the following problems.

1. A square has an area of 71 square units. What are its dimensions?

2. A triangle with two sides of the same length is called an **isosceles triangle**. If an isosceles right triangle has a hypotenuse of 12 centimeters, what is its area?

3. Use Figure 10.8 to solve the following.

   (a) Find the area of the square in Figure 10.8 using the Pythagorean theorem. (The grid is made up of unit squares.)

   (b) Check your result by counting the square units contained in the square.

4. A right triangle with legs of length a and b and a hypotenuse of length c can be used to form a square with sides of length $a + b$ by positioning four copies of the triangle, as shown in Figures 10.9 and 10.10.

   Explain why in this construction the four angles in the four-sided unshaded figure are all equal (Figure 10.10).

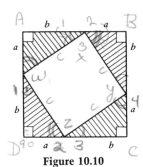

Figure 10.9

Figure 10.10

PROOF OF THE PYTHAGOREAN THEOREM

We want to prove now that the Pythagorean theorem is true. Remember that the theorem says, in a right triangle with legs of lengths a and b and hypotenuse of length c, that $c^2 = a^2 + b^2$. The argument uses principles of arithmetic as well as facts from geometry. Assume that we have a right triangle with legs of lengths a and b and hypotenuse of length c. Using four copies of the triangle in Figure 10.9 and positioning them as shown in Figure 10.10 we can form a square whose sides have length $a + b$. In problem 4 above, you saw that the four angles of the unshaded figure must all be equal. The four sides all have the same length c, so we can conclude that the unshaded figure is also a square.

There are two ways to write the areas of the four right triangles. The first is $4 \cdot \left(\frac{1}{2}ab\right)$ or $2ab$. The second is to view the area of the four right triangles as the area of the large square minus the area of the small square: $(a + b)^2 - c^2$. The two expressions must be equal.

$$(a + b)^2 - c^2 = 2ab \qquad \text{(Area of four triangles)}$$
$$a^2 + 2ab + b^2 - c^2 = 2ab \qquad \text{(Special product } (a + b)^2)$$
$$a^2 + b^2 - c^2 = 0 \qquad \text{(Subtract } 2ab \text{ from both sides)}$$
$$a^2 + b^2 = c^2 \qquad \text{(Add } c^2 \text{ to both sides)}$$

This is the result the Pythagorean theorem claims is true.

Although we will not give the argument in this book, it is also the case that, if we have three positive numbers a, b, c for which $c^2 = a^2 + b^2$, then there is a right triangle with sides of lengths a, b, and c. When three numbers a, b, and c satisfy the relationship $c^2 = a^2 + b^2$, they are called a **Pythagorean triple.**

10.3 SOLVING GENERAL QUADRATIC EQUATIONS

In the last section we saw that the special equation $x^2 = n$ for a positive number n has two solutions: $x = \sqrt{n}$ and $x\ -\sqrt{n}$. This is the simplest of the quadratic (degree 2) equations in one variable. In this section we discuss methods of solving more general quadratic equations.

There is a second special form of quadratic equation that is easily solved. Consider the equation $x^2 - 4x + 5 = 0$. The left side of this equation factors to give $(x - 5)(x + 1) = 0$. It is a basic property of arithmetic that when the product of two numbers is 0 at least one of the factors is 0. Thus, in this example, $x - 5 = 0$ or $x + 1 = 0$. In the first case we have $x = 5$; in the second case $x = -1$. Thus, 5 and -1 are two solutions to this equation. (Check them.) We say we have solved the equation by **factoring**. Here are two more examples where factoring gives solutions to quadratic equations.

EXAMPLE 1   Solve $x^2 + 3x = -2$.

$$x^2 + 3x + 2 = 0 \qquad \text{(Add 2 to both sides)}$$
$$(x + 2)(x + 1) = 0 \qquad \text{(Factor left-hand side)}$$

Either $x + 2 = 0$ or $x + 1 = 0$; thus, either

$$x = -2 \qquad \text{or} \qquad x = -1$$

EXAMPLE 2   Solve $-2x^2 + 7x - 3 = 0$.

$$(-2x + 1)(x - 3) = 0$$

Either $-2x + 1 = 0$ or $x - 3 = 0$; thus, either

$$x = \frac{1}{2} \qquad \text{or} \qquad x = 3$$

We observe that, if the quadratic expression can be factored, then factoring is a very useful technique for solving the equation. But there are many quadratic expressions that we know cannot be factored, and we must continue our search for a way to solve all quadratic equations.

Look carefully at the coefficients of quadratic expressions that can be written as squares of sums. For example,

$$x^2 + 8x + 16 = (x + 4)^2$$

$$x^2 + 14x + 49 = (x + 7)^2$$

$$x^2 + 2nx + n^2 = (x + n)^2$$

Observe in the examples that the numerical term on the left side is obtained by dividing the coefficient of x by 2 and squaring the result. A similar thing happens when a quadratic is the square of a difference, x minus a number.

$$x^2 - 8x + 16 = (x - 4)^2$$

$$x^2 - 14x + 49 = (x - 7)^2$$

$$x^2 - 2nx + n^2 = (x - n)^2$$

The last term on the left side is obtained by dividing the coefficient of x by 2 and squaring the result. This observation gives us a method for finding the solutions of any quadratic. The method is called **completing the square**. It is clumsy but it is reliable. Follow the steps in these examples to see how it works.

EXAMPLE 1 Solve $x^2 + 2x - 3 = 0$.

$$x^2 + 2x = 3 \qquad \text{(Add 3 to both sides)}$$
$$x^2 + 2x + 1 = 3 + 1 \qquad \text{(Divide the coefficient of } x \text{ by 2, square, and add to both sides)}$$
$$(x + 1)^2 = 4 \qquad \text{(Factor left side)}$$

Either $x + 1 = 2$ or $x + 1 = -2$, so either

$$x = 1 \qquad \text{or} \qquad x = -3$$

EXAMPLE 2 Solve $x^2 - 4x - 1 = 0$.

$$x^2 - 4x = 1 \qquad \text{(Add 1 to both sides)}$$
$$x^2 - 4x + 4 = 1 + 4 \qquad \text{(Divide the coefficient of } x \text{ by 2, square, and add to both sides)}$$
$$(x - 2)^2 = 5 \qquad \text{(Factor left side)}$$

Either $x - 2 = \sqrt{5}$ or $x - 2 = -\sqrt{5}$, so either

$$x = 2 + \sqrt{5} \quad \text{or} \quad x = 2 - \sqrt{5}$$

We know that the solutions to a quadratic equation must depend not on the symbol we use for the variable, but on the numbers that are coefficients in the equation. For example $x^2 - x - 4 = 0$ and $t^2 - t + 4 = 0$ have the same solutions. Therefore, we should be able to say in terms of coefficients a, b, and c what the solutions are for the equation $ax^2 + bx + c = 0$. We can use our technique of completing the square to derive the solutions of the general equation.

EXAMPLE 3   Solve $ax^2 + bx + c = 0$.

$$x^2 + \frac{b}{a}x + \frac{c}{a} = 0$$ (Divide both sides by a to get 1 as the coefficient of x^2 as in other examples)

$$x^2 + \frac{b}{a}x = -\frac{c}{a}$$ $\left(\text{Subtract } \frac{c}{a} \text{ from both sides}\right)$

$$x^2 + \frac{b}{a}x + \frac{b^2}{4a^2} = -\frac{c}{a} + \frac{b^2}{4a^2}$$ (Divide the coefficient of x by 2, square, and add to both sides)

$$\left(x + \frac{b}{2a}\right)^2 = \frac{b^2}{4a^2} - \frac{c}{a}$$ (Factor left side)

$$\left(x + \frac{b}{2a}\right)^2 = \frac{b^2 - 4ac}{4a^2}$$ (Combine terms on right side)

Either

$$x + \frac{b}{2a} = \sqrt{\frac{b^2 - 4ac}{4a^2}} \quad \text{or} \quad x + \frac{b}{2a} = -\sqrt{\frac{b^2 - 4ac}{4a^2}}$$

so either

$$x = -\frac{b}{2a} + \sqrt{\frac{b^2 - 4ac}{2a}} \quad \text{or} \quad x = -\frac{b}{2a} - \sqrt{\frac{b^2 - 4ac}{2a}}$$

As incredible as these solutions look, they are in fact extremely useful to us because we can now write the solutions to a quadratic equation just by knowing its coefficients. The simplified expressions $x = \dfrac{-b + \sqrt{b^2 - 4ac}}{2a}$ and $x = \dfrac{-b - \sqrt{b^2 - 4ac}}{2a}$ are called the **quadratic formula**. Here are some examples of how it can work.

EXAMPLE 4   Solve $x^2 - 4x - 1 = 0$.

Here $a = 1$, $b = -4$, and $c = -1$. Since the solutions are

$$x = \frac{-b + \sqrt{b^2 - 4ac}}{2a} \quad \text{and} \quad x = \frac{-b - \sqrt{b^2 - 4ac}}{2a}$$

we have

$$x = \frac{4 + \sqrt{16 + 4}}{2} \quad \text{and} \quad x = \frac{4 - \sqrt{16 + 4}}{2}$$

We can simplify to get

$$x = \frac{4 + \sqrt{20}}{2} = \frac{4 + 2\sqrt{5}}{2} = 2 + \sqrt{5}$$

$$\text{and} \quad x = \frac{4 - \sqrt{20}}{2} = \frac{4 - 2\sqrt{5}}{2} = 2 - \sqrt{5}$$

Notice that these solutions agree with the solutions we obtained to this equation when we used the method of completing the square on page 172.

EXAMPLE 5   Solve $4x^2 - 4x + 1 = 0$.

We know

$$x = \frac{-b + \sqrt{b^2 - 4ac}}{2a} \quad \text{and} \quad x = \frac{-b - \sqrt{b^2 - 4ac}}{2a}$$

where $a = 4$, $b = -4$, and $c = 1$. Thus,

$$x = \frac{4 + \sqrt{16 - 16}}{8} \quad \text{or} \quad x = \frac{4 - \sqrt{16 - 16}}{8}$$

Our solutions are $x = \frac{4 + 0}{8} = \frac{1}{2}$ and $x = \frac{4 - 0}{8} = \frac{1}{2}$. This equation has only the one solution $x = \frac{1}{2}$.

EXAMPLE 6   Solve $2x^2 + x + 3 = 0$.

Now $a = 2$, $b = 1$, and $c = 3$. Substitution into $x = \frac{-b + \sqrt{b^2 - 4ac}}{2a}$ gives $x = \frac{-1 + \sqrt{1 - 24}}{4}$ or $x = \frac{-1 + \sqrt{-23}}{4}$. However, there is no real number $\sqrt{-23}$ and hence no number $\frac{-1 + \sqrt{-23}}{4}$. When we evaluate $x = \frac{-b - \sqrt{b^2 - 4ac}}{2a}$ we have the same situation: an expression containing $\sqrt{-23}$. We conclude that the given equation has no real number solutions. (In later courses you may learn about number systems in which equations like this do have solutions.)

In Example 4 the equation has two different solutions, in Example 5 the equation has one solution, and in Example 6 no solutions

[handwritten notes:]
If discrimite = 0 → one root
if discrimite > 0 (positive #) = 2 roots
< 0 negative = 2 roots
$x = \dfrac{-b \pm \sqrt{b^2 - 4ac}}{2}$ will always solve problems

exist. The differences arise because in the first case $b^2 - 4ac$ is a positive number, in the second case $b^2 - 4ac$ is 0, and in the third case $b^2 - 4ac$ is a negative number. In the next section we explore the geometric meaning of these three possibilities for quadratic equations.

You can see that the quadratic formula gives a very efficient method of solving quadratic equations, so much so that you will probably want to commit it to memory. Although it is always possible to fall back on the method of completing the square, knowing the quadratic formula is a time saver.

Exercises 10-B

I. For each of the following, decide whether the equation has one, two, or no solutions by computing $b^2 - 4ac$.

1. $x^2 + 5x + 4 = 0$
2. $x^2 - 2x - 48 = 0$
3. $3x^2 - x - 4 = 0$
4. $2x^2 + 2x + 1 = 0$
5. $4x^2 - 49 = 0$
6. $5x^2 - 10x - 5 = 0$
7. $6x^2 + 13x - 5 = 0$
8. $x^2 + 4x - 117 = 0$
9. $4x^2 = 21x + 18$
10. $12x - 3 = 9x^2 + 1$ *$-9x^2 + 12x - 4$*
11. $x^2 + x + 1 = 0$
12. $12x^2 + 5x - 3 = 0$

II. Find the solutions for each of the equations in Part I above. You may factor if possible or use the quadratic formula. If there are no real number solutions, state this.

III. Solve the following problems.

1. A contractor wants to build a rectangular flower bed with an area of 200 square feet. One of the long sides of the bed will be the wall of a building and the other three sides will be fenced. He has 50 feet of fence available. What dimensions should the flower bed be?

2. One side of a rectangle is 5 inches longer than the other. The area of the rectangle is 84 square inches. Find the lengths of the sides.

3. Find two numbers whose sum is 5 and whose product is -36.

4. Find two consecutive numbers so that the sum of their squares is 145.

5. A concrete walk of uniform width extends around a rectangular lawn having dimensions of 20 by 80 feet. Find the width of the walk if the area of the walk is 864 square feet.

6. The hypotenuse of a right triangle is 13 inches. If one leg is 7 inches longer than the other, how long are the legs of the triangle?

7. A wire is stretched from the top of a fence to the top of a 20 foot vertical pole. If the fence and the pole are 30 feet apart and the wire is 34 feet long, find the height of the fence.

IV. Solve the following problems using graphs to explore the possibilities.

1. Sketch the graph of this quadratic equation in two variables: $y = x^2 - 2x - 3$. Be sure to plot many points. Where does the graph touch the horizontal axis? For what values of x is $0 = x^2 - 2x - 3$?

2. Do problem 1 using the equation $y = x^2 + x + 1$; and then using $y = x^2 - 4x + 4$.

3. Your graph of $y = x^2 + x + 1$ should look like Figure 10.11. On the same axis sketch the graph of $y = -x^2 - x - 1$.

4. Referring to problems 1 and 2, sketch the graphs of $y = -x^2 + 2x + 3$ and $y = -x^2 + 4x - 4$.

5. Suppose that the graph of $y = ax^2 + 2x + c$ contains the point $(-1, 2)$ and crosses the horizontal axis at $x = 2$. Find the numbers a and c. Does the graph cross the horizontal axis at a second point?

6. The following quadratic equations have related graphs. Graph each and then describe the relationship.

   (a) $y = x^2$
   (b) $y = (x + 2)^2$
   (c) $y = (x - 2)^2$
   (d) $y + 1 = x^2$
   (e) $y - 4 = x^2$
   (f) $y - 2 = (x - 1)^2$

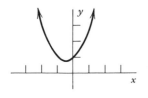

Figure 10.11

10.4   THE GEOMETRY OF QUADRATIC EQUATIONS

In Exercises 10-B you drew graphs of several quadratic equations of the form $y = ax^2 + bx + c$. In general, these graphs look like ⌣ or ⌢ . The graph of a quadratic equation is called a **parabola.** If the coefficient a is positive, then for large positive values of x and large negative values of x the values of y will be positive and large, and the parabola will open up like a cup ⌣ . But if the coefficient a is negative, then large positive and negative values of x will give negative y values, and the parabola will open down ⌢ . (In this case we might call it a "cap.")

The points on the graph of the equation $y = ax^2 + bx + c$ that lie on the horizontal axis have the second coordinate 0. Thus, when we look for the points where a graph crosses the horizontal axis, we look for x values that make $y = 0$ in the equation; that is, by asking where the graph of $y = ax^2 + bx + c$ crosses the horizontal axis is the same as asking what are the values of x that make $ax^2 + bx + c = 0$ (see Figure 10.12). The solutions to $ax^2 + bx + c = 0$ are the points on the horizontal axis where the graph of $y = ax^2 + bx + c$ crosses the axis.

Consider the example, $y = -x^2 + 5x - 4$. Even before graphing the parabola that this equation describes, we know that it will be a cap

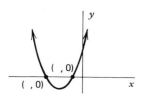

Figure 10.12

opening down because the coefficient of x^2 is negative. To find the points on the horizontal axis that are solutions to $y = -x^2 + 5x - 4$, we need the values for x that give $y = 0$; that is, $-x^2 + 5x - 4 = 0$. This equation is like those we solved in the previous section. We can use either factoring or the quadratic formula. In the quadratic formula we have $a = -1$, $b = 5$, and $c = -4$, so

$$x = \frac{-5 + \sqrt{25 - 16}}{-2} \qquad \text{or} \qquad x = \frac{-5 - \sqrt{25 - 16}}{-2}$$

Thus,

$$x = \frac{-5 + 3}{-2} = 1 \qquad \text{or} \qquad x = \frac{-5 - 3}{-2} = 4$$

We have two different solutions to the equation $-x^2 + 5x - 4 = 0$ because $25 - 16$ is a positive number. Now we have some very helpful information about the graph $y = -x^2 + 5x - 4$.

1. The graph is a parabola opening downward (a cap).

2. The graph crosses the horizontal axis at $x = 1$ and $x = 4$.

   We have, in fact, another piece of information because a parabola is a symmetric curve. (Look again at the parabolas in problem 6, Part IV, Exercises 10-B). In this example the symmetry requires that the highest point on the graph lie midway between $x = 1$ and $x = 4$. We find this value for x by taking the average $\frac{1 + 4}{2} = 2.5$. The corresponding y value is $y = -(2.5)^2 + 5(2.5) - 4$. Use your calculator to see that $y = 2.25$. Thus,

3. The highest point on the graph has coordinates (2.5, 2.25).

   Finally, it is easy to compute where the graph crosses the vertical axis. The x coordinate must be 0 for any point on the vertical axis, so $y = -x^2 + 5x - 4 = -0^2 + 5(0) - 4 = -4$.

4. The graph crosses the vertical axis at $y = -4$.

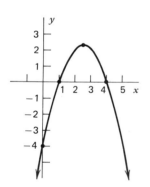

Figure 10.13

The graph consequently looks like Figure 10.13.

   We have seen that the number of solutions to an equation $0 = ax^2 + bx + c$ depends on the value of $b^2 - 4ac$. Specifically, here is the correspondence:

$$\text{2 solutions} \leftrightarrow b^2 - 4ac \text{ is positive}$$

$$\text{1 solution} \leftrightarrow b^2 - 4ac \text{ is 0}$$

$$\text{0 solutions} \leftrightarrow b^2 - 4ac \text{ is negative}$$

Now we also know that solutions to $0 = ax^2 + bx + c$ correspond to points where the graph of $y = ax^2 + bx + c$ crosses the horizontal axis. Thus, we can add a second correspondence to the one above:

Graph of $y = ax^2 + bx + c$ crosses the horizontal axis at	Equation $0 = ax^2 + bx + c$ has	Number $b^2 - 4ac$ is
2 points	2 solutions	positive
1 point	1 solution	0
0 points	0 solutions	negative

Thus, computing $b^2 - 4ac$ tells us how many times the graph of $y = ax^2 + bx + c$ crosses the horizontal axis; by solving $0 = ax^2 + bx + c$, we find out where.

Exercises 10-C

I. For each of the following equations, determine
  (a) Whether the graph crosses the horizontal axis at 0, 1, or 2 points.
  (b) Where the graph crosses the horizontal axis if it does.
  (c) The coordinates of the highest or lowest point on the graph.
  (d) Where the graph crosses the vertical axis.
  1. $y = x^2 + x - 2$ **4.** $y = 4x^2 - 4x + 1$
  2. $y = x^2 + 5x + 6$ **5.** $y = x^2 + 1$
  3. $y = 9 - x^2$

II. Suppose that $0 = ax^2 + bx + c$ has two solutions

$$x = \frac{-b + \sqrt{b^2 - 4ac}}{2a} \quad \text{and} \quad x = \frac{-b - \sqrt{b^2 - 4ac}}{2a}$$

with $b^2 - 4ac > 0$. These are the points on the horizontal axis where the graph of $y = ax^2 + bx + c$ crosses the axis. Find the point midway between these two points in terms of a, b, and c. This gives the x coordinate of the highest (or lowest) point on the graph of $y = ax^2 + bx + c$. Even in those cases where the graph of $y = ax^2 + bx + c$ does not cross the horizontal axis, the expression you have found is the x coordinate of the highest (or lowest) point on the graph.

10.5 FRACTIONAL EQUATIONS

Some problem situations give rise to equations that contain rational expressions. If these equations can be rewritten as quadratic equations, then they can be solved either by factoring or by using the quadratic formula. Here is such an example.

EXAMPLE 1   A boat that travels 12 miles per hour in still water takes 2 hours less time to go 45 miles downstream than to return the same distance upstream. What is the rate of the current?

We can summarize the information in this problem by making a table.

	Distance	Rate	Time
Downstream	45	$12 + r$	$\dfrac{45}{12 + r}$
Upstream	45	$12 - r$	$\dfrac{45}{12 - r}$

We assume that r is a positive number less than 12 because, if r were greater than 12, then $12 - r$ would be negative. Since the problem says the time going downstream is 2 hours less than the time going upstream, the equation $\dfrac{45}{12 + r} = \dfrac{45}{12 - r} - 2$ describes the problem. We need a method for solving this kind of equation. One thing we could do would be admit we don't like fractions in the equation. Remember that $(12 + r) \cdot \dfrac{45}{12 + r} = 45$ and $(12 - r) \cdot \dfrac{45}{12 - r} = 45$. Thus, if we multiply both sides of the equation by $(12 + r)(12 - r)$, we will have an equivalent equation with no fractions in it.

$$(12 + r)(12 - r)\left(\frac{45}{12 + r}\right)$$
$$= (12 + r)(12 - r)\left(\frac{45}{12 - r}\right) - 2(12 + r)(12 - r)$$
$$(12 - r)(45) = (12 + r)(45) - 2(144 - r^2)$$
$$540 - 45r = 540 + 45r - 288 + 2r^2$$
$$0 = 2r^2 + 90r - 288$$
$$0 = r^2 + 45r - 144$$

Now we have a quadratic equation equivalent to the original equation. We can solve it by factoring.

$$0 = r^2 + 45r - 144$$
$$0 = (r + 48)(r - 3)$$
$$r = -48 \quad \text{or} \quad r = 3$$

Since the rate of the current must be a positive number, we conclude the rate is 3 miles per hour.

Study carefully the solutions to these equations.

EXAMPLE 2   Solve $\dfrac{x + 1}{x} = \dfrac{x}{1}$.

$$\frac{x + 1}{x} = \frac{x}{1}$$

$$x + 1 = x^2 \qquad\qquad \text{(Multiply both sides by x)}$$

$$0 = x^2 - x - 1$$

Either $x = \dfrac{1 + \sqrt{1 + 4}}{2}$ or $x = \dfrac{1 - \sqrt{1 + 4}}{2}$. Since the rational expressions in the equation are defined for each of these numbers, we have two solutions: $x = \dfrac{1 + \sqrt{5}}{2}$ and $x = \dfrac{1 - \sqrt{5}}{2}$.

EXAMPLE 3   Solve $\dfrac{1}{x} + \dfrac{1}{x + 1} = \dfrac{5}{6}$.

$$\frac{1}{x} + \frac{1}{x + 1} = \frac{5}{6}$$

$$6(x + 1) + 6x = 5x(x + 1) \qquad \text{[Multiply both sides by } 6x(x + 1)]$$

$$6x + 6 + 6x = 5x^2 + 5x$$

$$0 = 5x^2 - 7x - 6$$

$$0 = (5x + 3)(x - 2)$$

Either $5x + 3 = 0$ or $x - 2 = 0$. Thus, $x = -\dfrac{3}{5}$ or $x = 2$. Both are solutions, and you can check this.

✳ The main idea behind this method of solving the problems is to choose a multiplier for both sides of the equation that contains as factors all the denominators appearing in the equation. We do need to restrict the values x can assume, however, so that no denominator in the original equation takes the value 0. The following example shows that a value can appear to be a solution and not be a solution.

EXAMPLE 4   Solve $\dfrac{x}{x - 5} = \dfrac{5}{x - 5} + 7$.

$$\frac{x}{x - 5} = \frac{5}{x - 5} + 7$$

$$x = 5 + 7(x - 5) \qquad \text{(Multiply both sides by } x - 5)$$
$$x = 5 + 7x - 35$$
$$30 = 6x$$
$$5 = x$$

But 5 is not a solution to the original equation because the rational expressions $\dfrac{x}{x-5}$ and $\dfrac{5}{x-5}$ are not defined for $x = 5$. We must conclude that this equation has no solutions.

One common problem situation that gives rise to an equation with fractional expressions is this type.

EXAMPLE 5   Art can paint his mother's house in 5 days. His brother can paint it in 7. If they work together, how long should it take them to paint the house?

Here it is helpful to analyze what happens in one day of work. (Assume that x represents the number of days it will take the men to paint the house working together.) In one day, Art paints $\dfrac{1}{5}$ of the house, his brother paints $\dfrac{1}{7}$, and together they paint $\dfrac{1}{x}$. Thus, $\dfrac{1}{5} + \dfrac{1}{7} = \dfrac{1}{x}$. We can solve as we have previously.

$$\frac{1}{5} + \frac{1}{7} = \frac{1}{x}$$
$$7x + 5x = 35 \qquad\qquad\qquad \text{(Multiply both sides}$$
$$\qquad\qquad\qquad\qquad\qquad\qquad \text{(by } 5 \cdot 7 \cdot x)$$
$$12x = 35$$
$$x = \frac{35}{12} \quad \text{or} \quad 2\frac{11}{12} \text{ days}$$

Exercises 10-D

I. Solve the following equations.

1. $\dfrac{5}{x} = \dfrac{15}{2}$

2. $\dfrac{x}{x-2} = \dfrac{3}{x-2}$

3. $x = \dfrac{-4}{x-4}$

4. $\dfrac{2}{x+6} = 2$

5. $\dfrac{t+3}{t-1} = \dfrac{t-1}{t+4}$

6. $\dfrac{x+3}{x-2} = \dfrac{x-2}{x+1}$

7. $\dfrac{3}{x+2} + \dfrac{4}{x-2} = 1$

8. $\dfrac{2x}{2x+3} = \dfrac{x-1}{x-3}$

9. $\dfrac{x}{x-2} = \dfrac{3x}{x-2} + 1$

10. $\dfrac{1}{x-2} + \dfrac{3}{x+1} = \dfrac{3}{x^2-x-2}$

11. $\dfrac{y}{y+2} = \dfrac{2y}{y+2} - 2$

12. $\dfrac{2}{x+1} - \dfrac{4}{x-1} = \dfrac{x}{x^2-1}$

13. $\dfrac{4}{x+5} - \dfrac{2}{x-3} = \dfrac{-16}{x^2+2x-15}$

II. Solve the following problems.

1. Sara and Lee paddle their canoe an average of 2 miles per hour in still water. They have finished a trip in a river where they went 15 miles upstream and then returned to their starting point after 16 hours of canoeing. What was the average rate of the stream?

2. The north pump can fill a swimming pool in 6 hours; the south pump can fill it in 4 hours. How long should it take to fill the pool if both pumps are working?

3. The north pump can fill a (different) swimming pool in 6 hours. The time it takes for the south pump to fill it is 1 hour longer than the two pumps together. Find the time it takes the two pumps together.

4. A motorist is driving between two towns 200 miles apart. If she would increase her speed by 10 miles per hour, she would save 1 hour. How fast is she going?

5. An airplane flies 1275 miles in the same time it takes a boat to go 25 miles. If the plane goes 500 miles per hour faster than the boat, find the rate of each.

REVIEW PROBLEMS

1. Perform the indicated operations and simplify the following.
   (a) $(4\sqrt{a} - \sqrt{b})(4\sqrt{a} + \sqrt{b})$
   (b) $\sqrt{27a^3} - a\sqrt{12a}$
   (c) $(\sqrt{x^4y^5})^2$

2. Find the distance between the points $(-1, 4)$ and $(3, -2)$.

3. A square has a diagonal of length 5 inches. What is the length of a side?

4. Solve the following equations.
   (a) $5x^2 + 13x - 6 = 0$
   (b) $x^2 + 2x = x - 1$
   (c) $\dfrac{5}{x-2} - \dfrac{3}{x-3} = \dfrac{1}{x-2}$
   (d) $\dfrac{x+1}{x} = \dfrac{x}{2} + 1$

5. For each of the following determine the points where the graph of the equation crosses the horizontal axis and the coordinates of the highest or lowest point on the graph. Then sketch the graph.
   (a) $y = -x^2 + 2x + 3$
   (b) $y = 2x^2 + 12x + 15$

6. A support wire is stretched from the top of a 120-foot tower to the ground and is anchored 50 feet from the base of the tower. How long is the wire?

7. Terry walks 2 miles to the bus stop and then takes the bus 15 miles to work. The average speed of the bus is 10 times her walking speed. If the total trip takes $1\frac{1}{6}$ hours, what is her walking speed and the speed of the bus?

8. Ted and Joan Williams have replaced a picture window that is 34 inches wider than it is tall. They have been charged for 5040 square inches of glass. What are the dimensions of their window?

Completing a squares (rewrite form so you can find values of x

$x^2 + 10x + 7$ (does not factor)

$x^2 + 10x + 7 = 0$

$x^2 + 10x = -7$ ½ of b cofficient + squared add to both side

$x^2 + 10x + 25 = -7 + 25$

$(x+5)^2 = 18$ take square root of both sides

$x + 5 = \pm\sqrt{18}$

$x = -5 \pm \sqrt{\frac{18}{9 \cdot 2}}$

$x = -5 \pm 3\sqrt{2}$

CHAPTER 11

VARIATION AND NEGATIVE EXPONENTS

11.1 VARIATION

In the preceding chapters we have examined many problem situations in which two or more variables are related to each other. Recall the following four examples.

1. A motorist travels 300 miles. If the rate at which he travels is known, the time the trip takes can be computed (Chapter 1).

If r is the rate and t is the time, then

$$t = \frac{300}{r}$$

2. A rectangle has an area of 144. If the length is known, the width can be computed (Chapter 1).

If ℓ is the length and w is the width, then

$$w = \frac{144}{\ell}$$

3. Two travelers start traveling in opposite directions from the same point at the same time, one at 15 miles per hour and the other at 25 miles per hour. If we know how long they have been traveling, we can compute the distance between them (Chapter 1).

If t is the time traveled and D is the distance between them, then
$$D = 40 \cdot t$$

4. A phone company charges 9¢ per local call. If the number of calls placed is known, the bill can be computed (Chapter 1).

If N is the number of calls and B is the amount of the bill, then
$$B = 0.09 \cdot N$$

Examining these four problems, you will notice several similarities and differences. In each of the problems the equation used for the solution states that one variable can be determined by multiplying or dividing a number (called a constant) by the other variable. We express this by saying that one variable varies with the other. In problems 3 and 4 the constant is multiplied by a variable; in these cases both variables get larger (or smaller) together. For example, the larger the number of calls, the larger the phone bill. In such cases variation is **direct,** and we say that the phone bill **varies directly** with the number of calls or that the phone bill is directly proportional to the number of calls. In problem 3 we would say that the distance **varies directly** with the time.

In problems 1 and 2 the variation is not direct. In problem 1 a large (fast) rate results in a small time, and a small (slow) rate results in a large time. In this case we say that the time **varies inversely** with the rate. In problem 2 the width **varies inversely** with the length of the rectangle. We now give a formal definition of these two types of variation.

If x and y are variables, then **y varies directly with x** provided there is a nonzero number k such that $y = k \cdot x$.

If x and y are variables, then **y varies inversely with x** provided there is a nonzero number k such that $y = \dfrac{k}{x}$.

In both cases above, the number k is called the **constant of variation.**

We consider the concept of variation further by examining three other problems.

EXAMPLE 1 An electric bill from Friendy Edison varies directly with the number of kilowatt hours of electricity used. The Jones' used 450 kilowatt hours and paid $15 last month. This month they used 520 kilowatt hours of electricity. What should they expect their bill to be?

Let B stand for the amount billed and h stand for the kilowatt hours used. Since B varies directly with h, $B = k \cdot h$ for some number k. Now we can use last month's data.

$$15 = k \cdot 450$$

$$\frac{15}{450} = k \qquad \text{(Divide both sides by 450)}$$

$$\frac{1}{30} = k$$

Thus, the constant of variation is $k = \dfrac{1}{30}$. This month the equation $B = k \cdot h$ becomes

$$B = \frac{1}{30} \cdot 520$$

$$B = \frac{52}{3}$$

$$B = \$17.32$$

In Example 1, knowing that two quantities varied directly and knowing their values at one time enabled us to compute the constant of variation. We then used this constant for computations at other times. We examine this procedure in another situation.

EXAMPLE 2   The distance required to stop a car varies directly with the square of the speed at the time the brakes are applied. Jim's car needs 187.5 feet to stop when it is going 50 miles per hour. How many feet are needed when the car is going 65 miles per hour?

Let D be the stopping distance and S be the speed. Then if k is the constant of variation, we have $D = k \cdot S^2$. Using the information given,

$$187.5 = k \cdot (50)^2$$
$$\frac{187.5}{2500} = k$$
$$0.071 = k$$

Therefore, when the car is traveling 65 miles per hour, the stopping distance is given by

$$D = 0.071 \cdot (65)^2$$
$$D = 299.975 \text{ feet}$$

A similar procedure can be used for solving problems involving inverse variations.

EXAMPLE 3   A group of office workers are "chipping in" to buy a retirement present for a co-worker. The amount each person must chip in varies inversely with the number of people participating. If 20 people participate, each must give $5. How much must each person give if 32 people participate?

Let N be the number of participants and C be the cost per person. Then $C = \dfrac{k}{N}$ where k is the constant of variation. Using the data from the problem, we obtain

$$\$5 = \frac{k}{20}$$
$$100 = k$$

Thus, when $N = 32$, $C = \dfrac{100}{32} = 3.125$. Each person should give $3.13.

Note that in this problem k is in fact the cost of the present.

Our procedure in each of these problems has been the same.

1. Determine the type of variation from the problem.
2. Write an equation involving the constant of variation.
3. Use data from the problem to find the constant.
4. Use the constant in the equation to answer the question of the problem.

We need to examine one other type of variation before turning to the exercises. This is the idea of **joint variation.** We say a variable x varies jointly with two or more other variables when x varies directly with their product. For example, the area of a triangle varies jointly with its base and height. The area formula $A = \frac{1}{2} b \cdot h$ demonstrates that area varies directly with the product of b and h; the constant of variation is $\frac{1}{2}$. Notice that if either of the variables b or h is held constant, the area varies directly with the other variable. A formal definition of joint variation follows.

If x, y, and z are variables, then z varies jointly with x and y provided there is a nonzero number k such that $z = k \cdot x \cdot y$.

In many scientific applications of variation a quantity varies directly with one variable and inversely with another variable. We consider two examples of this type. First, suppose a family is planning a vacation trip. Among other things they need to decide how much money they can spend on their trip and what type of accommodations they wish. The number of days they can afford to spend will vary directly with the amount of money they have available and inversely with the daily cost of their lodging and food. If M stands for the money available, D for the average daily costs, and N for the number of days, this problem can be written as $N = \frac{M}{D}$. In this example, the constant of variation is the number 1.

The kinetic theory of matter provides us with one of the many examples of complex variation in physics. This theory says that the volume V filled by a gas varies directly with the number N of molecules of gas present, directly with the temperature T, and inversely with the pressure P at which the gas is stored. Written algebraically, $V = k\frac{N \cdot T}{P}$; k is a constant of variation that depends on the nature of the particular gas.

Exercises 11-A

I. Find the constant of variation in each of the following.
1. A varies directly with B, and $B = 110$ when $A = 14$.
2. x varies directly with y^3, and $y = 2$ when $x = 16$.
3. E varies directly with t^2, and $E = 40$ when $t = 5$.

4. J varies inversely with p, and $J = 10$ when $p = 4$.

5. R varies inversely with s^2, and $R = 100$ when $s = 4$.

6. P varies inversely with q^4, and $p = 10$ when $q = 3$.

7. A varies jointly with c and d, and $A = 75$ when $c = 4$ and $d = 3$.

8. x varies directly with y and inversely with z, and $x = 4$ when $y = 3$ and $z = 2$.

9. t varies jointly with u^2 and v, and $t = 2$ when $u = 5$ and $v = 4$.

10. z varies directly with p and inversely with r^3, and $z = 1$ when $r = 2$ and $p = 5$.

II. Describe the type of variation and identify the constant of variation in each of the following equations.

1. $A = \pi r^2$

2. $C = 2\pi r$

3. $D = rt$

4. $t = \dfrac{d}{r}$

5. $y = \dfrac{1}{x^2}$

6. $y = (x + 2)^2$

7. $r = 100t^2$

8. $\dfrac{x}{y} = 4y$

9. $2xy = 9$

10. $xyz = 1$

III. Solve the following problems.

1. The distance required for a truck to stop varies directly with the square of its speed. How much distance is required for the truck to stop from a speed of 60 miles per hour if 40 feet are required when it is traveling at 20 miles per hour?

2. A manufacturer makes boxes of uniform height; the volume of the boxes varies jointly with the length and width. What is the volume of a box 1.5 feet long and 2 feet wide if the volume of a box 3 feet long and 3 feet wide is 18 cubic feet.

3. The time it takes a jogger to complete 10 laps around the high school track varies inversely with his speed. Mike runs at 5.2 miles per hour and needs 30 minutes to run 10 laps. How long will it take Bill to run 10 laps at 7.1 miles per hour?

The idea of variation has many uses in science. The following problems illustrate some of these uses.

4. The distance an object falls varies directly with the square of the time it takes to fall. An object dropped from a building 400 feet high falls in 5 seconds. How long will it take an object to fall to the ground from a mile high skyscraper?

5. Sound intensity (loudness) varies inversely with the square of the distance from the sound. A rock band measures 100 decibels at 50 feet. What does it measure at 5 feet?

6. The amount a spring stretches varies directly with the force applied. If a 5-pound weight stretches the spring 3 inches, how far will a 2-pound force stretch the spring?

11.2 USING NEGATIVE NUMBERS AS EXPONENTS

In our study of variation in the last section, and in much of our work throughout this course, we have seen that multiplication and division are closely related to each other. Although it is useful to distinguish two types of variation, direct and inverse variation, as we have above, you should observe that the following statements are equivalent.

A varies inversely with x^2

A varies directly with $\dfrac{1}{x^2}$

This is one example of the close relationship between concepts defined in terms of division and concepts defined in terms of multiplication. In this section we extend the multiplicative concept of exponents to division; in doing so we introduce negative exponents.

Recall the following information about exponents from Chapter 2.

Definition: If n is a whole number and b is any number, then b^n denotes the product of n numbers each of which is b. (For example, $2^3 = 2 \cdot 2 \cdot 2 = 8$.)

Rule 1: $b^n \cdot b^m = b^{n+m}$.

Rule 2: $(a \cdot b)^n = a^n \cdot b^n$.

Rule 3: $(b^n)^m = b^{n \cdot m}$.

Our definition of positive exponents and our three rules for computing with exponents all were developed in terms of multiplication. Just as we were concerned that the principles of arithmetic be maintained when we expanded our number system to include negative numbers and fractions, we will be careful that our extension to negative exponents is consistent with the three rules of exponents. Consider the following sequence of powers of 2.

Power of 2: 2^1 2^2 2^3 2^4 2^5 2^6 2^7 . . .
Value: 2   4   8   16 32 64 128 . . .

As we proceed in the development of this sequence, at each step we observe that two things are happening; the number at each step in the sequence is multiplied by 2, and the exponent at each step is increased by 1. Suppose we begin at some point in this sequence, say $2^3 = 8$, and reverse the process: at each step we divide the value by 2, or equivalently **decrease** the exponent by 1. We get the following sequence.

Power of 2: 2^3 2^2 2^1 2^0 2^{-1} 2^{-2} 2^{-3} . . .

Value: 8   4   2   1 $\frac{1}{2}$ $\frac{1}{4}$ $\frac{1}{8}$ . . .

Thus, reversing the process of raising the base 2 to positive powers suggests that 2^0 has the value 1 and that $2^{-n} = \frac{1}{2^n}$. An analogous demonstration would show a similar relation for any base. Thus, we state two definitions.

$b^0 = 1$ for any nonzero base b.

$b^{-n} = \frac{1}{b^n}$ for any nonzero base b and any positive integer n.

In many books these definitions of zero and negative exponents are generated by applying the laws of exponents to the symbols. In this approach we ask, for example, "What numbers *should* the symbols 5^0 and 5^{-2} equal?" and find solutions by applying the laws of exponents. See if you can follow these arguments.

1. What number should 5^0 be?

      Consider $5^3 \cdot 5^0 = 5^{3+0}$ (Rule 1 for exponents)

             $5^3 \cdot 5^0 = 5^3$ (Addition)

                  $5^0 = 1$ (Divide both sides by 5^3)

Notice that what we have done here is solve for the value of 5^0 and thus determined that $5^0 = 1$. A similar procedure would work for 2^0, 3^0, or any base with a zero exponent. Try it!

2. What number should 5^{-2} be?

      Consider $5^3 \cdot 5^{-2} = 5^{3+(-2)}$ (Rule 1 for exponents)

           $5^3 \cdot 5^{-2} = 5$ (Addition)

               $5^{-2} = \frac{5}{5^3}$ (Divide both sides by 5^3)

               $5^{-2} = \frac{1}{5^2}$ (Reduce fractions)

Again, we have solved an equation using Rule 1 for exponents and found $5^{-2} = \frac{1}{5^2}$.

SIMPLIFYING EXPRESSIONS CONTAINING EXPONENTS

We can now use the rules of exponents together with our definitions of zero and negative exponents to simplify many expressions involving exponents. In general, we say that an expression is **simplified** when it has been reduced to a form in which each base appears only once and in which there are no negative or zero exponents. Examine the following examples carefully, and then use them as a guide for doing the exercises.

EXAMPLE 1   Simplify $x \cdot x^{-3}$.

$$x \cdot x^{-3} = x^{1+(-3)} \qquad \text{(Rule 1 for exponents)}$$
$$= x^{-2} \qquad \text{(Addition)}$$
$$= \frac{1}{x^2} \qquad \text{(Definition of negative exponents)}$$

EXAMPLE 2   Simplify $y \cdot (2y^{-1})^2$.

$$y \cdot (2y^{-1})^2 = y \cdot 2^2 \cdot (y^{-1})^2 \qquad \text{(Rule 2 for exponents)}$$
$$= y \cdot 2^2 \cdot y^{-2} \qquad \text{(Rule 3 for exponents)}$$
$$= 4y \cdot y^{-2} \qquad \text{(Commutative principle)}$$
$$= 4y^{1+(-2)} \qquad \text{(Rule 1 for exponents)}$$
$$= 4y^{-1} \qquad \text{(Addition)}$$
$$= \frac{4}{y} \qquad \text{(Definition of negative exponents)}$$

Two consequences of rules 1, 2, and 3 for exponents are often stated as separate rules for computing with exponents. They refer to exponents in the quotients of expressions.

RULE 4 FOR EXPONENTS   **(Finding the quotient of powers of the same base)**

$$\frac{b^n}{b^m} = b^{n-m}$$

RULE 5 FOR EXPONENTS   **(Finding a power of a quotient)**

$$\left(\frac{a}{b}\right)^n = \frac{a^n}{b^n}$$

Look at the following example to see how these rules are used in simplifying expressions that contain exponents.

EXAMPLE 3   Simplify $\left(\dfrac{xy^{-1}}{x^2z}\right)^{-2}$.

$$\left(\frac{xy^{-1}}{x^2z}\right)^{-2} = \left(\frac{x^{-1}y^{-1}}{z}\right)^{-2}$$ (Rule 4 for exponents: x terms)

$$= \frac{(x^{-1}y^{-1})^{-2}}{z^{-2}}$$ (Rule 5 for exponents)

$$= \frac{x^2y^2}{\cdot z^{-2}}$$ (Rule 2 for exponents)

$$= \frac{x^2y^2}{\dfrac{1}{z^2}} \quad {}= x^2y^2 \cdot \frac{z^2}{1}$$ (Definition of negative exponents)

$$= x^2y^2z^2$$ (Rules for operations on fractions)

One final observation about negative exponents is useful in computation. It is a direct consequence of the meaning of a negative exponents. See if you can state this result in words.

EXAMPLE 4   Simplify $\dfrac{x^{-2}}{y^{-3}}$.

$$\frac{x^{-2}}{y^{-3}} = \frac{\dfrac{1}{x^2}}{\dfrac{1}{y^3}}$$

$$= \frac{1}{x^2} \div \frac{1}{y^3}$$

$$= \frac{1}{x^2} \cdot \frac{y^3}{1}$$

$$= \frac{y^3}{x^2}$$

Thus,

$$\frac{x^{-2}}{y^{-3}} = \frac{y^3}{x^2}$$

Exercises 11-B

I. Compute the following.

   1. 8^0

   2. 8^{-2}

   3. 8^{-3}

   4. $-5^0 = -1$

always will be -1

5. $(-5)^0 =$ 1   *always will be 1*

6. -5^{-2}

7. $(-5)^{-2}$

8. $3 \cdot 2^{-1}$

9. $6 \cdot 6^{-1}$

10. $2 \cdot 3^{-2}$

II. Simplify the following expressions.

1. $x^{-3} \cdot x^5$

2. $x^3 \cdot x^{-5}$

3. $x^{-6} \cdot x^2$

4. $x^6 \cdot x^{-2} \cdot (-x)^0$

5. $x^4 \cdot x^{-2} \cdot x^1$

6. $t^3 \cdot t^{-6} \cdot t^{-1}$

7. $u^{-3} \cdot u^6 \cdot u^{-1} \cdot u^0$

8. $a^3 \cdot (ab)^4 \cdot b^{-6}$

9. $\left(\dfrac{a}{b}\right)^{-3}$

10. $(a^{-4})^{-3}$

11. $(a^4)^{-3}$

12. $(-2y^2)^{-3}$

13. $(a^{-2} \cdot b^{-1})^{-2}$

14. $(ab^2)^{-1} \cdot (a^{-1}b^{-2})^{-1}$

15. $(a^2b^5)^{-2} \cdot a^{12}b^2$

16. $x^{-2}y^7(x^{-2}y^{-3})^{-3}$

17. $(-3x^{-2})^{-4} \cdot (3x^2)^4$

18. $(-2x^2)^{-3} \cdot (3x^{-2})^3$

19. $(-a \cdot b^2 \cdot c^{-3})^{-4}$

20. $[2(-2x^2)^{-3}]^{-1}$

21. $\left(\dfrac{xy^{-3}}{x^2y^2}\right)^{-1}$

22. $\left(\dfrac{2x^2}{-2y^{-2}}\right)^{-3}$

23. $\left(\dfrac{xy^{-3}}{z} \cdot \dfrac{z^{-2} \cdot x^2}{y^{-2}}\right)^{-1}$

24. $\dfrac{(-3x)^{-2}}{y^{-3}} \cdot \dfrac{(3x)^2}{y^{-3}}$

25. $\left[\left(\dfrac{x}{y}\right)^{-2}\right]^{-1}$

26. $\dfrac{a^5b^{-2}c^{-3}}{a^6b^{-3}c^{-3}}$

27. $a^2 \cdot \left(\dfrac{a^4b}{a^{-3}b^2}\right)^{-3} \cdot b^{-2}$

28. $x^2y^2(x^{-2} + y^{-2})$

29. $\left(\dfrac{2}{y} + 5\right)^{-1}$

30. $\dfrac{1}{\left(\dfrac{x}{y} - 2\right)^{-2}}$

31. $\dfrac{\dfrac{1}{x^{-1}}}{(x+1)^{-2}}$

32. $\dfrac{\dfrac{1}{x} + 5}{x^{-3}}$

REVIEW PROBLEMS

1. Find the constant of variation in each of the following.

  (a) x varies directly with r, and $r = 77$ when $x = 11$.

  (b) t varies inversely with p, and $t = 15$ when $p = 3$.

(c) r varies inversely with w^3, and $r = 8$ when $w = 2$.

2. Describe the type of variation and identify the constant of variation in each of the following.

(a) $xy^2 = 5$ (b) $t = \dfrac{2w}{z}$

3. The number of people that can be served from a soup tureen varies inversely with the size of the servings. If 6-ounce cups are used, 24 people can be served. How many people can be served if 8-ounce cups are used?

4. Compute the following.

(a) 7^{-2} (b) $(-2)^{-3}$

(c) $4 \cdot 5^{-2}$

5. Simplify each of the following.

(a) $x^5 \cdot x^{-6}$ (c) $(x^{-2}y)^{-3}$

(b) $(ab)^{-2} \cdot (a^3b^2)$ (d) $\left(\dfrac{xy^3z^{-2}}{x^{-3}y^2z}\right)^{-2}$

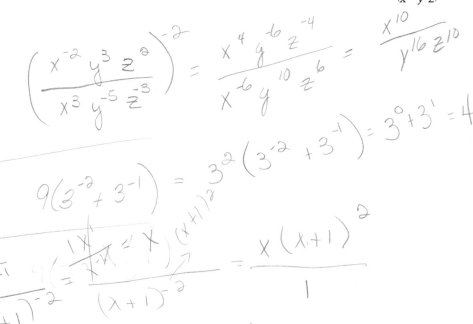

CHAPTER 12

CONIC SECTIONS, FUNCTIONS, AND FUNCTIONAL NOTATION

12.1   PARABOLAS

There are four different kinds of curves that can be formed when a plane intersects a double cone. These are illustrated in Figure 12.1.

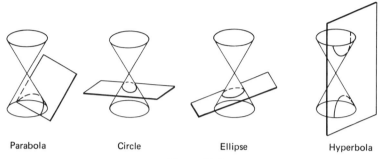

Parabola   Circle   Ellipse   Hyperbola

Figure 12.1

A **parabola** is formed when the plane is parallel to one of the lines in the surface of the cone. A plane parallel to the base of a cone cuts a **circle.** If the plane that cuts a circle is tilted, an **ellipse** will result. If the plane cuts both the upper and lower halves of the double cone, a **hyperbola** is formed. The four curves are called **conic sections** because they can be cut from a double cone.

In 200 B.C. the Greek mathematician Apollonius wrote a series of eight books which systematically described the properties of conic sections. We study some of these properties in this chapter. A parabola, for example, is the collection of points that are the same distance from a point F and a line ℓ; that is, if the point F and the line ℓ are given, then P is on the parabola if the distance from P to F equals the perpendicular distance from P to ℓ (Figure 12.2).

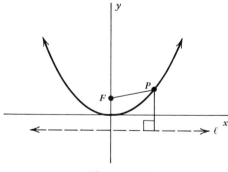

Figure 12.2

If a ball is thrown or a rocket shot near the surface of the earth, its path will be a parabola. The cable of a suspension bridge, if the load is uniformly distributed, assumes the shape of a parabola. Parabolic mirrors are commonly used in headlights because they reflect light rays parallel to the ground. The parabola is a curve that is commonly used in applications of mathematics and, therefore, it is an important curve to study.

Descartes' development of analytic geometry in the seventeenth century made it possible to write equations that describe the conic sections. Recall that parabolas are described by quadratic equations. In the equation $y = ax^2 + bx + c$, if the coefficient a is a positive number, the parabola opens up like a cup; if the coefficient a is negative, the parabola opens down like a cap. There are many families of parabolas; parabolas within the same family are the same shape but are positioned differently with respect to the coordinate axes.

EXAMPLE Compare the graphs of $y = x^2$, $y = (x - 1)^2$, $y = (x + 3)^2$, $(y - 1) = x^2$, $(y + 3) = x^2$, and $(y + 3) = (x - 1)^2$.

If we plot enough points to sketch these six graphs, we get the results illustrated in Figures 12.3 to 12.8.

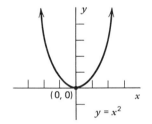

Figure 12.3

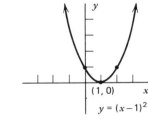

Figure 12.4

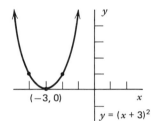

Figure 12.5

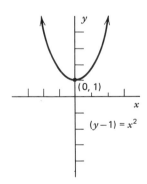

Figure 12.6

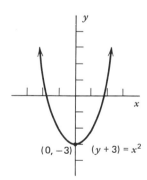

Figure 12.7

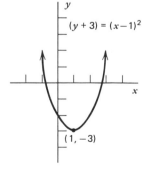

Figure 12.8

All of the parabolas in the example have the same shape. In each case the position of the vertex can be read from the equation describing the parabola. Observe that the parabola described by $y = x^2$ has its vertex at $(0, 0)$, and the parabola described by $(y + 3) = (x - 1)^2$ has its vertex at $(1, -3)$. In fact, the graph of the equation $(y - k) = (x - h)^2$ is a parabola with a vertex at (h, k). For example, the graph of $(y - 2) = (x + 3)^2$ has its vertex at $(-3, 2)$.

Reversing the roles of the two variables in an equation reverses the position of the graph with respect to the two axes. Consider the equation $x = y^2$ (or $y^2 = x$). For each positive value of x, there are two y values.

x	1	1	4	4	5	5
y	1	-1	2	-2	$\sqrt{5}$	$-\sqrt{5}$

If $x = 0$, then $y = 0$; there is no y value that corresponds to a negative value for x. Here are the graphs of $y^2 = x$ and $y^2 = -x$ (Figures 12.9 and 12.10). Notice that the parabola opens to the left when the coefficient of x is negative, and that the parabola opens to the right when the coefficient of x is positive.

Replacing y by $y + 3$ and x by $x - 1$ has the effect of moving the vertex to the point $(1, -3)$ as we can see in Figures 12.11 and 12.12.

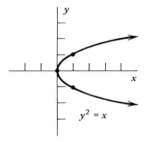

$y^2 = x$

Figure 12.9

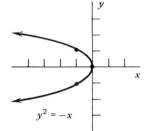

$y^2 = -x$

Figure 12.10

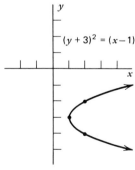

$(y + 3)^2 = (x - 1)$

Figure 12.11

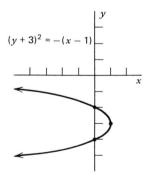

$(y + 3)^2 = -(x - 1)$

Figure 12.12

However, not all parabolas look like the ones in Figures 12.9 to 12.12. The parabolas in Figures 12.13 to 12.16, for example, are not in the same family as those above. Notice that the coefficient of the squared term determines whether the parabola is narrow or wide.

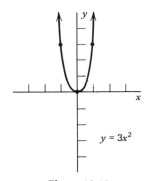

Figure 12.13

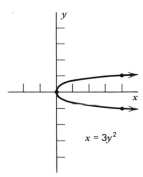

Figure 12.14

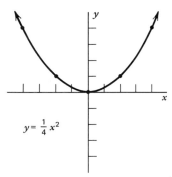

Figure 12.15

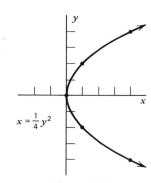

Figure 12.16

Exercises 12-A

I. Graph each of the following parabolas.

1. $4x^2 - y = 0$ 3. $4y^2 - x = 0$
2. $y - 2x^2 = 0$ 4. $x - 2y^2 = 0$

II. Solve the following problems.

1. The distance between the towers of a suspension bridge is 800 feet. The lowest point on the cables is 120 feet below the points of support. Assume that the cables have a parabolic shape. Taking the lowest point on one of them as the origin of a coordinate system, write an equation describing the shape of a cable.

2. If an object is thrown straight upward vertically from the ground at an initial speed of 64 feet per second, its distance d above the earth t

seconds after it is thrown is given by the equation $d = 64t - 16t^2$.

(a) Graph the equation $d = 64t - 16t^2$.

(b) What is the greatest height the object reaches? How long does it take to reach this height?

(c) How many seconds after launch does the object return to earth?

12.2   CIRCLES

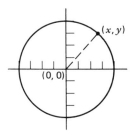

Figure 12.17

A second conic section is the circle. A **circle** can be described as the set of points at a given distance from a fixed point. The fixed point is called the **center** of the circle; the fixed distance is called the **radius.**

To write the equation for a circle, we must use the formula for computing the distance between two points in terms of their coordinates: the distance between (a, b) and (c, d) is $\sqrt{(a - c)^2 + (b - d)^2}$. (See Chapter 10, page 170, and Figure 12.17.) If we want the equation of the circle with radius 4 and center (0, 0), we express the fact that the distance between a point (x, y) on the circle and the center (0, 0) equals 4.

$$\sqrt{(x - 0)^2 + (y - 0)^2} = 4$$

$$\sqrt{x^2 + y^2} = 4$$

$$x^2 + y^2 = 16$$

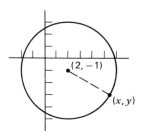

Figure 12.18

Writing the equation of a circle with the center at $(2, -1)$ and a radius of 4 is similar (Figure 12.18). We describe algebraically the fact that the distance between a point (x, y) on the circle and the center $(2, -1)$ equals 4.

$$\sqrt{(x - 2)^2 + [y - (-1)]^2} = 4$$

$$\sqrt{(x - 2)^2 + (y + 1)^2} = 4$$

$$(x - 2)^2 + (y + 1)^2 = 16$$

Both of the circles above have a radius of 4, but they have different centers. The equation $x^2 + y^2 = 16$ describes a circle with a center at (0, 0) and a radius of 4. The equation $(x - 2)^2 + (y + 1)^2 = 16$ describes a circle with its center at $(2, -1)$ and a radius of 4. In general, a circle with a center at (h, k) and a radius of r has the equation

$$(x - h)^2 + (y - k)^2 = r^2$$

If an equation has this form, it is easy to recognize that it describes a circle and to read both the center and the radius of the circle directly from the equation. However, an equation describing a circle may be in an equivalent but different form. For example, all of the following equations describe a circle with a center at $(-3, 2)$ and a radius of 4.

$$(x + 3)^2 + (y - 2)^2 = 16$$

$$x^2 + 6x + 9 + y^2 - 4y + 4 = 16$$

$$x^2 + 6x + y^2 - 4y + 13 = 16$$

$$x^2 + 6x + y^2 - 4y = 3$$

If we were given the equation $x^2 + 6x + y^2 - 4y = 3$, we might want to reverse the steps above and put it in the form $(x + 3)^2 + (y - 2)^2 = 16$ so that we could recognize it as a circle with the center $(-3, 2)$ and the radius 4. The process we use to do this is the same as completing the square, which we used to derive the quadratic formula. We must complete the square in the variable x and also in the variable y.

$$x^2 + 6x + y^2 - 4y = 3$$

$x^2 + 6x + 9 + y^2 - 4y = 3 + 9$ (Divide the coefficient of x by 2, square, add to both sides)

$x^2 + 6x + 9 + y^2 - 4y + 4 = 3 + 9 + 4$ (Divide the coefficient of y by 2, square, add to both sides)

$$(x + 3)^2 + (y - 2)^2 = 16$$

Exercises 12-B

I. Write the equation of a circle with the following center and radius.
  1. Center $(0, 0)$, radius 7   **3.** Center $(2, -9)$, radius 3
  2. Center $(0, 0)$, radius $\sqrt{5}$ **4.** Center $(-1, -4)$, radius 11

II. Find the center and radius of each circle whose equation is given. Then draw the graph.
  1. $(x + 1)^2 + (y - 2)^2 = 5$ **5.** $2x^2 + 2y^2 = 8$
  2. $(x - 12)^2 + (y - 10)^2 = 15$ **6.** $x^2 + y^2 - 8x + 10y - 5 = 0$
  3. $x^2 + y^2 - 9 = 0$ **7.** $2x^2 - 4x + 2y^2 + 12y - 12 = 0$
  4. $x^2 + y^2 - y = 0.75$

12.3   ELLIPSES

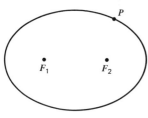

Figure 12.19

An **ellipse** is a curve that looks like a picture of a football with its ends rounded. Actually its shape is determined by two points that are called the **focus points** of the ellipse. The sum of the distance from a point P on the ellipse to F_1 plus the distance from P to F_2 is the same for any point P on the ellipse (Figure 12.19). One way to draw an ellipse is to take a fixed length of string and attach the ends with two thumb tacks (Figure 12.20). A pencil holding the string taut will draw an ellipse because the sum of the distances from the pencil to the two tacks will always be the length of the string. (Try it!)

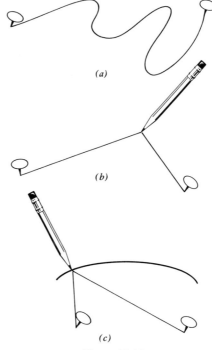

(a)

(b)

(c)

Figure 12.20

The fixed distance (or the length of the string) in an ellipse is related to the longest distance across the ellipse. Assume the longest distance across the ellipse is $2a$. Then the length of the string can be seen by placing the pencil at one end so that the string goes from F_1 to the pencil and from the pencil back to F_2; the distance is equal to $2a$ (Figure 12.21). If the pencil is moved as shown in Figure 12.22 so that the distance from F_1 to the pencil is the same as the distance from the pencil to F_2, each of these distances must be a because the sum is $2a$.

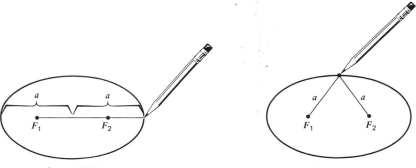

Figure 12.21   Figure 12.22

The point midway between the two focus points is called the center of the ellipse. If we join the pencil point to the center and the center to a focus point, a right triangle is formed. In Figure 12.23 the lengths of the legs of the right triangle are labeled b and c. In this triangle the hypotenuse is a, and we know the Pythagorean theorem says that

$$a^2 = b^2 + c^2$$

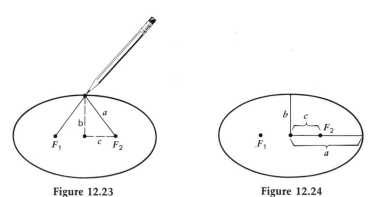

Figure 12.23   Figure 12.24

There are three important distances measured from the center of an ellipse. They are related by the equation $a^2 = b^2 + c^2$ (Figure 12.24).

We can write an equation for an ellipse from the information that the sum of the distances from a point on the ellipse to the two focus points is $2a$. Take for example the ellipse with its center at the origin, as shown in Figure 12.25. Assume the focus points have coordinates $(-c, 0)$ and $(c, 0)$. In this example $a = 4$ and $b = 3$. If (x, y) represents a point on the ellipse, then

$$\sqrt{[x - (-c)]^2 + (y - 0)^2} + \sqrt{(x - c)^2 + (y - 0)^2} = 2(4)$$

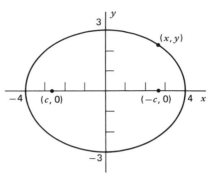

Figure 12.25

This is the basic equation but it can be simplified in several steps as follows.

$$\sqrt{(x + c)^2 + y^2} + \sqrt{(x - c)^2 + y^2} = 8$$
$$\sqrt{(x + c)^2 + y^2} = 8 - \sqrt{(x - c)^2 + y^2}$$

$$(x + c)^2 + y^2 = 64 - 16\sqrt{(x - c)^2 + y^2} \qquad \text{(Square both sides)}$$
$$\quad + (x - c)^2 + y^2$$

$$4\sqrt{(x - c)^2 + y^2} = 16 - cx \qquad \text{(Simplify)}$$
$$16(x - c)^2 + 16y^2 = 256 - 32cx + c^2x^2 \qquad \text{(Square both sides)}$$
$$x^2(16 - c^2) + 16y^2 = 16(16 - c^2) \qquad \text{(Simplify)}$$

Now it is important to remember that $a^2 = b^2 + c^2$ where $a = 4$ and $b = 3$. Thus, $16 = 9 + c^2$. Using the fact that $16 - c^2 = 9$, we can simplify the equation another step.

$$x^2 \cdot 9 + 16y^2 = 16 \cdot 9$$

$$\frac{x^2}{16} + \frac{y^2}{9} = 1 \qquad \text{(Divide both sides by } 16 \cdot 9\text{)}$$

$$\frac{x^2}{4^2} + \frac{y^2}{3^2} = 1$$

An interesting thing has happened in the simplification of the original equation. The coordinates of the focus points have disappeared. The important numbers in the equation are the distance $a = 4$ from the center to a vertex and the distance $b = 3$, the shortest distance from the center to the boundary. In general, the equation for an ellipse with a center at the origin and focus points on the x axis has the form

$$\frac{x^2}{a^2} + \frac{y^2}{b^2} = 1$$

where a is the distance from the center to a vertex and b is the shortest distance from the center to the boundary.

If we move the ellipse so that it has its center at $(-3, 2)$, the new ellipse will have the equation $\dfrac{(x + 3)^2}{4^2} + \dfrac{(y - 2)^2}{3^2} = 1$ (Figure 12.26).

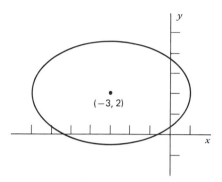

Figure 12.26

An equation could describe an ellipse but be in a form different than the equation above. An example is $x^2 + 4y^2 + 6x - 8y + 9 = 0$. We can complete the square in both variables like we did in the case of a circle and then read the important numbers for graphing from the equation. For example,

$$x^2 + 4y^2 + 6x - 8y + 9 = 0$$

$$x^2 + 6x + 4(y^2 - 2y) = -9$$

$$x^2 + 6x + 9 + 4(y^2 - 2y + 1) = -9 + 9 + 4$$

$$(x + 3)^2 + 4(y - 1)^2 = 4$$

$$\frac{(x + 3)^2}{4} + \frac{(y - 1)^2}{1} = 1$$

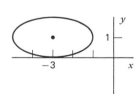

Figure 12.27

This equation describes an ellipse with the center at $(-3, 1)$ and $a = 2$ and $b = 1$ (Figure 12.27).

The equation of an ellipse may contain a larger number in the denominator of the y^2 term than in the denominator of the x^2 term. In our notation the number a then is under y^2 and the number b is under x^2. An example is $\dfrac{(x + 3)^2}{1} + \dfrac{(y - 1)^2}{4} = 1$. When this happens, the ellipse is "longer" in the vertical direction than in the horizontal direction. Compare Figure 12.28 to Figure 12.27. We are studying only conics that are "parallel" to the coordinate axes because their equations are relatively easy to describe. However, you should recognize that conics could be positioned anywhere in the plane.

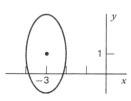

Figure 12.28

Even though Apollonius fully described the properties of the conics in 200 B.C., it was not until 1609 that Johannes Kepler used the ancient theory to make the brilliant observation that the orbits of planets are elliptical and not circles on circles, as had been believed by the Greeks. This application of mathematics was very slow in coming but very important in the understanding of the universe.

Exercises 12-C

I. For each of the following ellipses give the values of the constants a, b, and c. Write the coordinates of the center and the focus points. Finally, write the equation of the ellipse.

1.

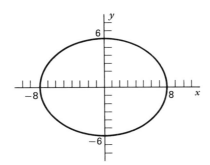

2.

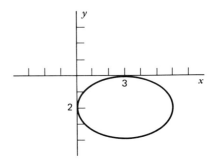

3.

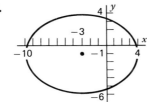

4.

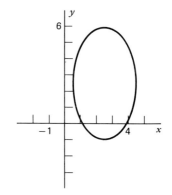

II. Each of the following equations describes an ellipse. Complete the square in x and y if necessary and write the equations in the form $\dfrac{(x - h)^2}{a^2} + \dfrac{(y - k)^2}{b^2} = 1$; then graph the equation.

1. $16x^2 + 25y^2 = 400$

2. $16x^2 + 9y^2 = 144$

3. $4x^2 + 24x + y^2 - 2y + 33 = 0$

4. $4x^2 - 8x + 9y^2 + 36y + 4 = 0$

12.4 HYPERBOLAS

A hyperbola, like an ellipse, is determined by two focus points. The **difference of the distances** from a point on the hyperbola to the two focus points is the same for any point on the hyperbola. In Figure 12.29 the distance from P to F_2 minus the distance from P to F_1 equals the distance from Q to F_1 minus the distance from Q to F_2.

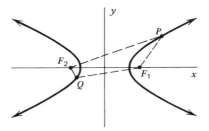

Figure 12.29

We can develop only a few of the characteristics of the hyperbola here. Our principal interests are in the equations that describe hyperbolas and the shapes of their graphs. The two parts of a hyperbola are called its **branches.** Each extends indefinitely and crosses the line joining the foci at a vertex (Figure 12.30). The point midway between the vertices is called the **center.** The further out you go on each branch, the closer you get to one of the two intersecting lines that acts as a guideline for the curve. These lines are called **asymptotes.** Asymptotes are not part of the hyperbola; they are, however, helpful in graphing a hyperbola.

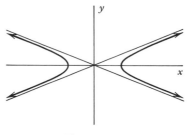

Figure 12.30

In Figure 12.31 the center of the hyperbola is shown at the origin. Three distances are important in writing the equation of a hyperbola. The distance a is the distance from the center to a vertex; the distance c is the distance from the center to a focus point; the distance b is the distance from a vertex to an asymptote along a line perpendicular to the line that runs through the focus points. If we draw a circle of radius c and center it at the center of the hyperbola, we can see the relationship between the three numbers a, b, and c (Figure 12.32). In a hyperbola,

$$c^2 = a^2 + b^2$$

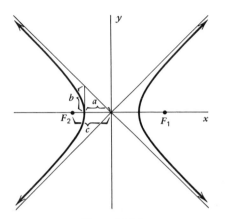

Figure 12.31

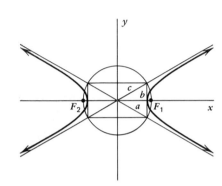

Figure 12.32

The equation of a hyperbola is a consequence of the fact that the difference of the distances from a point on the hyperbola to the two focus points is the same for any point on the hyperbola. We can see what this distance is by considering the vertex point V on the hyperbola (Figure 12.33). The distance from V to F_2 is $c + a$. The distance from V to F_1 is $c - a$. The difference is $c + a - (c - a) = c + a - c + a = 2a$. Thus, if (x, y) is any point on the hyperbola (Figure 12.34) the distance from (x, y) to F_1 minus the distance from (x, y) to F_2 must be $2a$.

$$\sqrt{(x - c)^2 + y^2} - \sqrt{(x + c)^2 + y^2} = 2a$$

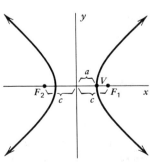

Figure 12.33

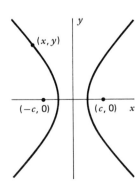

Figure 12.34

This equation simplifies after several steps of computation to the form

$$\frac{x^2}{a^2} - \frac{y^2}{b^2} = 1 \qquad \text{where } c^2 = a^2 + b^2$$

It is interesting to compare an ellipse and a hyperbola whose equations contain the same numbers a and b.

EXAMPLE On the same coordinate axes graph the following equations.

1. $\dfrac{x^2}{4} + \dfrac{y^2}{9} = 1$

2. $\dfrac{x^2}{4} - \dfrac{y^2}{9} = 1$

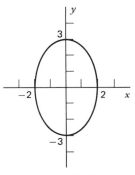

Figure 12.35

Equation 1 describes an ellipse centered at the origin. It cuts the x axis at 2 and -2, and cuts the y axis at 3 and -3 (Figure 12.35). Equation 2 describes a hyperbola centered at the origin. The numbers 2 and 3 determine the asymptotes; we can draw a rectangle and display the asymptotes as its diagonals (Figure 12.36). The vertices of the hyperbola are on the x axis at 2 and -2; the asymptotes guide the extremes of the hyperbola. Thus, we can obtain the final sketch in Figure 12.37. Note that the ellipse lies inside the rectangle and the hyperbola lies outside the rectangle.

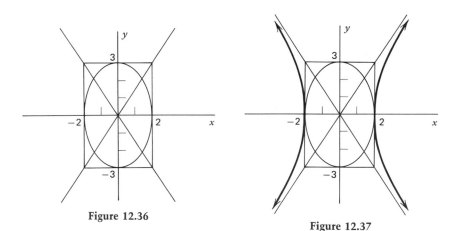

Figure 12.36

Figure 12.37

If the x and y variables are interchanged in the equation of a hyperbola, the graph of the new equation is a hyperbola intersecting the y axis rather than the x axis. The equation $\dfrac{y^2}{4} - \dfrac{x^2}{9} = 1$ has

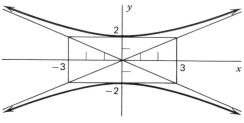

Figure 12.38

the graph shown in Figure 12.38.

Moving the hyperbola so that the center is at a point other than the origin changes the equation of the hyperbola in the same way that we have observed for the other conics. The equation $\dfrac{(y-2)^2}{4} - \dfrac{(x+3)^2}{9} = 1$ has a hyperbola for its graph, shaped like the one in Figure 12.38, but with a center at $(-3, 2)$.

To get an equation whose graph is a hyperbola into the form where we can read the coordinates of the center as well as the numbers a and b that determine the asymptotes, we again use the procedure of completing the square in both variables. In fact, we often cannot look at an equation and tell whether its graph is a circle, an ellipse, or a hyperbola until we have completed the square in both variables and changed its form.

EXAMPLE   Graph the equation $9x^2 - 4y^2 - 54x - 40y - 55 = 0$.

We complete the square in both variables in this way.

$$9x^2 - 4y^2 - 54x - 40y - 55 = 0$$
$$9(x^2 - 6x + \quad) - 4(y^2 + 10y + \quad) = 55$$
$$9(x^2 - 6x + 9) - 4(y^2 + 10y + 25) = 55 + 81 - 100$$
$$9(x - 3)^2 - 4(y + 5)^2 = 36$$
$$\frac{(x - 3)^2}{4} - \frac{(y + 5)^2}{9} = 1$$

Now we can see that the graph is a hyperbola with vertices on a line parallel to the x axis, with the center at $(3, -5)$, and with $a = 2$ and $b = 3$ (Figure 12.39).

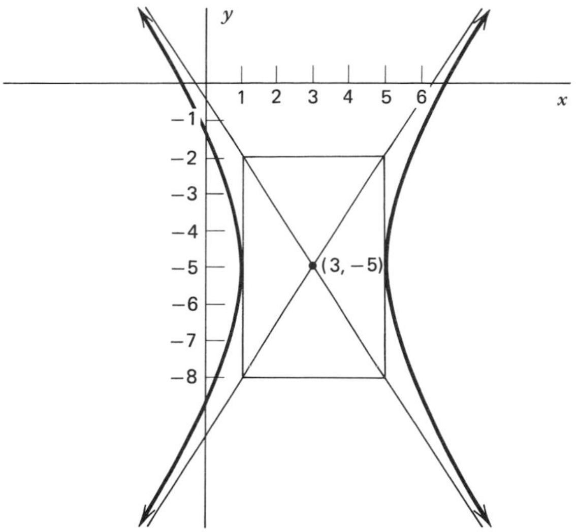

Figure 12.39

Exercises 12-D

I. Decide for each of the following equations whether the graph is an ellipse or a hyperbola. Then draw the graph.

1. $\dfrac{x^2}{4} + \dfrac{y^2}{16} = 1$

8. $\dfrac{(x+3)^2}{36} + \dfrac{(y-2)^2}{25} = 1$

2. $\dfrac{x^2}{25} - \dfrac{y^2}{9} = 1$

9. $x^2 - 12x + 3y^2 + 12y + 39 = 0$

3. $\dfrac{y^2}{25} - \dfrac{x^2}{9} = 1$

10. $25x^2 - 9y^2 - 100x - 72y - 269 = 0$

4. $\dfrac{x^2}{100} + \dfrac{y^2}{144} = 1$

11. $x^2 - 2x + y^2 - 2y - 6 = 0$

5. $\dfrac{x^2}{36} + \dfrac{y^2}{36} = 1$

12. $y^2 - 5x^2 + 20x = 50$

6. $\dfrac{(x-5)^2}{25} + \dfrac{y^2}{4} = 1$

13. $9x^2 + 4y^2 - 90x - 24y = -153$

7. $\dfrac{(x-2)^2}{9} - \dfrac{(y-1)^2}{16} = 1$

II. Solve the following problems.

1. Write the equations for the two asymptotes of the hyperbola with the equation

(a) $\dfrac{x^2}{25} - \dfrac{y^2}{16} = 1$

(b) $\dfrac{y^2}{64} - \dfrac{x^2}{36} = 1$

(c) $\dfrac{(x + 6)^2}{36} - \dfrac{(y - 3)^2}{9} = 1$

2. Write the equation of the hyperbola that passes through $(2, 0)$ and has asymptotes with the equations $x - 2y = 0$ and $x + 2y = 0$.

3. Write the equation of the hyperbola that passes through $(4, 2)$ and has asymptotes with the equations $y = 2x$ and $y = -2x + 4$.

12.5   FUNCTIONS AND FUNCTIONAL NOTATION

There are many situations in daily living where two events are associated or where one depends on the other. Each of the following statements describes a situation in which exactly one number is associated to each object or person in a collection of objects or persons.

1. To each toy in a toy store, assign its price.
2. To each third-grade child, assign his or her height.
3. To each resident of Clearwater, assign his or her house (apartment) number.

When we have a correspondence that assigns exactly one number to each object in a collection of objects, we call the correspondence a **function.**

During our study of mathematics in this book we have graphed a large number of equations. Some graphs have the property that for each value that the x variable can assume there is exactly one corresponding y value. Other graphs do not have this property. When an equation in the variables x and y has the property that there is exactly one y value corresponding to each x value, we say the equation describes a **function of x.** Thus, y is a function of x if to each value of x there is assigned one value of y.

EXAMPLE 1   The equation $y = x^2$ describes a function.

The graph of $y = x^2$ is a parabola that opens up (Figure 12.40). Each

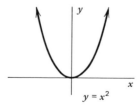

$y = x^2$

Figure 12.40

value for x determines exactly one value for y. This fact can be seen both from the equation and from its graph.

EXAMPLE 2 The equation $x^2 + y^2 = 4$ does not describe a function.

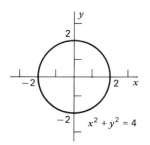

Figure 12.41

The equation $x^2 + y^2 = 4$ has a circle for its graph, a circle of radius 2 with its center at the origin (Figure 12.41). Observe that for each x value between -2 and 2 there are two corresponding y values. For example, when $x = 0$, y assumes both the values 2 and -2. When $x = 1$, y assumes the two values $\sqrt{3}$ and $-\sqrt{3}$. Thus, the equation $x^2 + y^2 = 4$ does not describe a function.

The graph of an equation contains enough information to determine if the equation describes a function or not. The graph of a function has only one point on the graph corresponding to each point on the x axis. If there is a point on the x axis that has more than one corresponding point on a graph, we can conclude we do not have the graph of a function. Compare Figures 12.42 and 12.43.

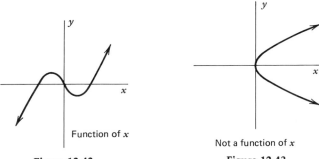

Figure 12.42 | Figure 12.43

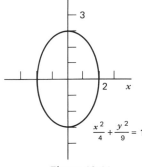

Figure 12.44

Circles, ellipses, and hyperbolas generally are not graphs of functions. However, it may be possible to describe parts of these graphs as functions. Consider the graph of $\frac{x^2}{4} + \frac{y^2}{9} = 1$, an ellipse with its center at the origin (Figure 12.44). When the equation is rewritten as $y^2 = 9\left(1 - \frac{x^2}{4}\right)$, we can see why two y values, one positive and one negative, correspond to each x value (except $x = 2$ and $x = -2$). This occurs because y^2 rather than y is in the equation. We can break the graph into two parts, each of which is described by a function. Examine Figures 12.45 and 12.46.

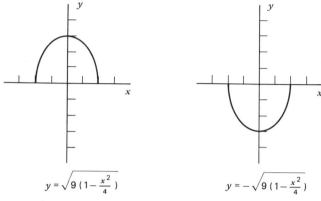

$$y = \sqrt{9\left(1-\frac{x^2}{4}\right)}$$

Figure 12.45

$$y = -\sqrt{9\left(1-\frac{x^2}{4}\right)}$$

Figure 12.46

FUNCTIONAL NOTATION

Since functions associate only one y value to each x value, it is possible to talk about the **one** y value that corresponds to a given x value. For example, the function $y = 2x + 3$ associates the y value 5 to the x value 1, the y value 3 to the x value 0, and the y value -1 to the x value -2. One way to show which y value corresponds to a given x value is to denote the function with a single letter, say f, and write

$f(1) = 5$	The function associates to the x value 1 the y value 5.
$f(0) = 3$	The function associates to the x value 0 the y value 5.
$f(-2) = -1$	The function associates the x value -2 the y value -1.
$f(x) = 2x + 3$	The function associates to any number x the number $2x + 3$.

Usually, we read the phrase $f(1) = 5$ with somewhat fewer words than we used above. More typically we might say, "the value of the function when $x = 1$ is 5," or, even more briefly, "f of 1 is 5."

For a second example, consider the function described by $g(x) = x^2 - x$. Here the function is denoted by the letter g and we are told that the y value associated with any x value is computed as $x^2 - x$. For example,

$$g(3) = 3^2 - 3 \qquad = 6$$
$$g(1) = 1^2 - 1 \qquad = 0$$
$$g(0) = 0^2 - 0 \qquad = 0$$
$$g(-2) = (-2)^2 - (-2) = 6$$

We do not use this kind of notation for equations that are not functions because the value of y for some x is not just one number.

Exercises 12-E

I. For the three functions

$$f(x) = (x + 1)^2 - x$$
$$g(x) = x - 2(x + 3)$$
$$h(x) = x - 5x^2 - 2$$

find the following values.

1. $f(1)$ **6.** $g(t)$

2. $f(0)$ **7.** $h\left(\dfrac{1}{2}\right)$

3. $f(-1)$ **8.** $h(0)$
4. $g(-3)$ **9.** $h(2t)$
5. $g(10)$ **10.** $f(3) - g(1)$

II. Solve the following problems.

1. Consider the function $f(x) = x^2 - x$. Find $f(4)$, $f(2)$, $f\left(\dfrac{1}{2}\right)$, $f(-1)$, and $f(-3)$. Use these values together with the ones computed in the narrative above to draw a graph of the function f.

2. Graph the function $s(x) = \sqrt{x}$ using values of x that are either positive or 0. Why do we exclude negative values of x for this function?

3. Draw a graph of the equation $y^2 = x$. Does this equation describe a function? Compare this graph to the one in problem 2.

4. Which of the following are graphs of functions? Explain your answers.

(a)
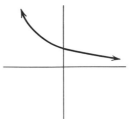

(b)
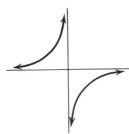

(c)
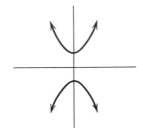

(d)
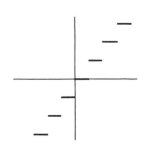

5. Is every line a graph of a function?

6. If $p(x)$ is a polynomial, is $y = p(x)$ a function? Graph $y = p(x)$ for
   (a) $p(x) = x^3$ (c) $p(x) = x^3 - 4x$
   (b) $p(x) = x^4$

7. The graph that follows is a hyperbola with the equation $\dfrac{y^2}{16} - \dfrac{x^2}{4} = 1$.

   Give two functions, each of which has one branch of the hyperbola for its graph.

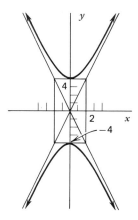

8. The graph that follows is a hyperbola with the equation $\dfrac{x^2}{16} - \dfrac{y^2}{4} = 1$.

   Break the graph into four parts, each of which can be described by a function.

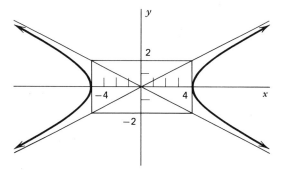

REVIEW PROBLEMS

1. Write an equation for each of the following conic sections. For each give the coordinates of the center and the focus points.

(a)

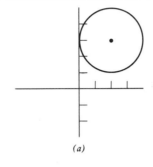

(a)

(b)

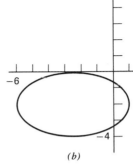

(b)

(c)

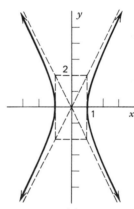

(c)

2. The graph of $y = -x^2$ follows. Sketch the graph of $y - 1 = -(x + 2)^2$.

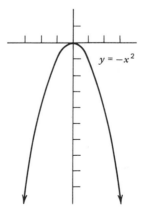

3. Graph each of the following equations.

(a) $x^2 + y^2 - 6x + 2y - 6 = 0$ (c) $9x^2 + 4y^2 - 18x - 27 = 0$

(b) $x - y^2 = 0$ (d) $9x^2 - 4y^2 - 18x - 27 = 0$

4. Graph the hyperbola $y^2 - \dfrac{x^2}{4} = 1$. Give two functions, each of which has one branch of the hyperbola for its graph.

5. If an object is thrown straight upward from the ground at an initial speed of 80 feet per second, its distance (in feet) t seconds after it is thrown is given by the function

$$d(t) = 80t - 16t^2$$

Graph the function to find the maximum height the object reaches. When does it come back to the ground?

PREFACE TO APPENDIXES

There are several topics and procedures in elementary mathematics that adult students may or may not remember, depending on whether they have needed to use the topics since they studied them in school. The purpose of these appendixes is to provide both a reference and an opportunity for independent review of these topics.

Each appendix is accompanied by a brief Pretest to help you determine what you remember about a given topic. Before beginning work on an appendix, take the Pretest; then compare your answers with those given and read the advice on how to plan your study of the topic.

APPENDIX A

AREA AND PERIMETER

PRETEST

1. Find the area of the rectangle in Figure A.1.
2. Find the perimeter of the rectangle in Figure A.1.
3. Find the area of the shaded triangle in Figure A.2.

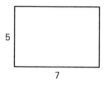

Figure A.1

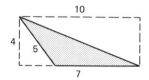

Figure A.2

4. Find the perimeter of the shaded area in Figure A.3.
5. Find the area of the trapezoid in Figure A.4.

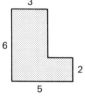

Figure A.3

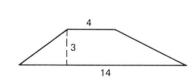

Figure A.4

ANSWERS TO PRETEST

1. 35
2. 24
3. 14

4. 22
5. 27

If you missed more than one of problems 1, 3, and 5, work through Section A.1.

If you missed more than one of problems 2 and 4, work through Section A.2.

Area and perimeter provide two different ways of looking at the size of geometric figures. The **area** of a figure is a measure of the surface of the figure; if a person is interested in buying enough paint to cover a wall or enough grass seed to cover a lawn, the area of the surface is the important measurement. On the other hand, if the same person were interested in buying molding to go around the wall or fencing to enclose the lawn, the important measurement would be the **perimeter.**

A.1 AREA OF POLYGONS

When we talk about area, we assume that there is a square that has a side length of 1 (sometimes called a **unit square**). The area of a figure tells us the number of unit squares needed to fill the figure completely without overlapping. For many figures it is almost impossible to see directly how many unit squares fill the figure. However, for a rectangle whose sides have natural number lengths, this is not difficult. Consider, for example, a rectangle with a base of length 5 and a height of 3. As Figure A.5 shows, this rectangle is covered by 15 nonoverlapping unit squares. We can count the number of unit squares in a rectangle or we can compute the number by multiplying the length of the base b by the height h. For a rectangle, the area is $b \cdot h$. Since a square is a rectangle with the length of its base equal to its height, the area of a square that has side length s is $s \cdot s$ or s^2.

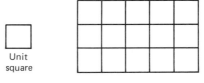

Unit
square

Figure A.5

These formulas apply to squares and rectangles even if the lengths of the sides are not whole numbers. Consider, for example the rectangle with base 4 and height 1.5, as shown in Figure A.6. Here the rectangle is covered by 4 unit squares and 4 other pieces; each piece is half of a unit square, so together the pieces are equivalent to 2 unit squares. In all, 6 unit squares cover the rectangle. You can use your calculator to see that $4 \times (1.5)$ does, in fact, equal 6.

Unit
square

Figure A.6

Knowing how to compute the area of a rectangle helps us see how to compute the areas of other closed figures that have line segments for their sides; these figures are called **polygons.** The assumption we make is that a figure can be cut apart and the area of the figure will equal the sum of the areas of the parts.

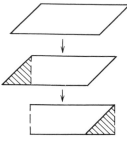

Figure A.7

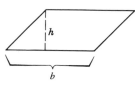

Figure A.8

AREA OF PARALLELOGRAMS

A *parallelogram* is a four-sided polygon in which the opposite sides are parallel. Opposite sides of a parallelogram have the same length. If you cut a right triangle off one end of a parallelogram and move it to the other end, you will form a rectangle (Figure A.7). Thus, the area of the parallelogram is computed by taking the base b times the height h. It is important to observe how the height of a parallelogram is measured. Take a look at Figure A.8.

Observing that the area of a parallelogram is $b \cdot h$ means we do not need to try to fit unit squares into a parallelogram and then count them to find its area. We can measure the height and base of the parallelogram and compute their product to get the area.

AREA OF TRIANGLES

To see how to compute the area of a triangle, draw lines through two of the vertices of the triangle parallel to the opposite sides (Figure A.9).

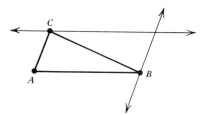

Figure A.9

Here we started with a triangle with vertices A, B, and C and drew a line through C parallel to $\overline{AB}$ and a line through B parallel to $\overline{AC}$. In doing this we have constructed a parallelogram, half of which is the triangle. Thus, the triangle has area $\frac{1}{2}$ times the area of the parallelogram or $\frac{1}{2}bh$ (Figure A.10). It is important to realize that the height

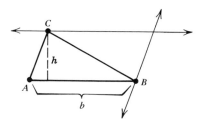

Figure A.10

of a triangle, like the height of a parallelogram, is measured from a vertex to the opposite side along a line perpendicular to the opposite side. A line on which the height is measured need not always lie inside the triangle. Figure A.11 illustrates two examples.

Figure A.11

AREA OF TRAPEZOIDS

A **trapezoid** is a four-sided polygon with at least one pair of opposite sides parallel. A trapezoid can be cut into two triangles and its area written as the sum of the areas of the triangles (Figure A.12).

Figure A.12

To write the areas of the two triangles we must identify their heights and the lengths of their bases. We pick the parallel sides of the trapezoid as bases of the triangles, and let c and b represent their lengths. An important observation is that the two triangles have the same height h (Figure A.13).

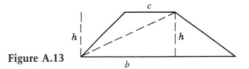

Figure A.13

One triangle has area $\frac{1}{2}hc$ (turn the page upside down to see this one), and the other has area $\frac{1}{2}hb$. Thus, the trapezoid has an area of $\frac{1}{2}hc + \frac{1}{2}hb$. The distributive property permits us to rewrite this expression as $\frac{1}{2}h(c + b)$. Another way to write the expression is $\frac{1}{2}(c + b)h$. In this form we see that the area of a trapezoid is the average of the two base lengths times the height.

AREA OF OTHER POLYGONS

We have been able to write concise expressions for the areas of several simple polygons as follows.

Area of rectangle:	bh
Area of parallelogram:	bh
Area of triangle:	$\frac{1}{2}bh$
Area of trapezoid:	$\frac{1}{2}(b + c)h$

Figure A.14

Of course, you must remember what the symbols represent in these expressions and, if you do, these formulas are useful in computing the areas for these polygons.

It is not always possible to write concise expressions for the areas of polygons with many sides. However, any polygon can be cut into triangles, and the area then computed as the sum of areas of triangles (Figures A.14 and A.15). Usually in computing the areas of figures that are not polygons we would need more advanced techniques than we are developing in this book. Therefore, our area examples will mostly involve polygons.

Figure A.15

Exercises A-1

I. Find the areas of each of the polygons that follow.

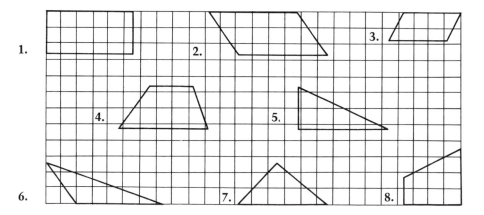

II. Find the area of each of the shaded regions in the following figures.

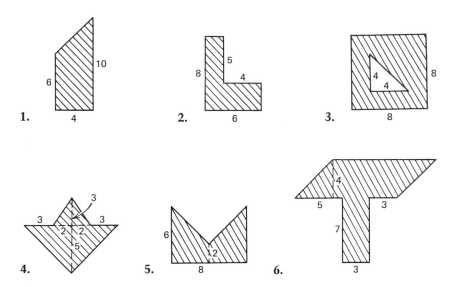

1.   **2.**   **3.**

4.   **5.**   **6.**

III. Solve the following problems.

   1. Explain why the following three triangles all have the same area.

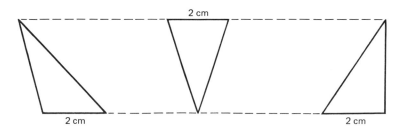

2 cm

2 cm   2 cm

   2. Use the area of a rectangle to explain the distributive property for multiplication with respect to subtraction. Use the equation

$$a \cdot (b - c) = a \cdot b - a \cdot c$$

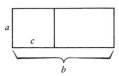

3. Use the area of a rectangle to explain why for any two numbers a and b,

$$(a + b)^2 = a^2 + 2ab + b^2$$

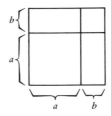

(This is a fact you will want to remember.)

A.2   PERIMETER OF POLYGONS

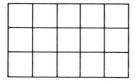

Figure A.16

The perimeter of a polygon (or of any geometric figure) is the distance around the figure. Return to the rectangle on page 225.

The unit of length we use to measure the perimeter of this rectangle is the side of the unit square used to measure the area (Figure A.16). Notice that the vertical sides of the rectangle are each 3 units long and the horizontal sides are each 5 units long. The perimeter is the sum of the lengths of the four sides: $3 + 3 + 5 + 5$, or $2 \cdot 3 + 2 \cdot 5 = 16$. The perimeter of any rectangle with base b and height h is $P = 2 \cdot b + 2 \cdot h$. This formula applies even when b and h are not natural numbers.

In general, the method for finding the perimeter of a polygon is to find the lengths of each of the sides and add together all these lengths. If the lengths are all given to you, this is an easy task. Sometimes, however, you will need to reason geometrically in order to find a perimeter. Consider this problem.

EXAMPLE   Find the perimeter of Figure A.17.

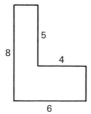

Figure A.17

The lengths of two of the sides are not given; we label them a and b (Figure A.18).

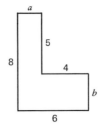

Figure A.18

Now, $a + 4 = 6$. To see this, cut the figure into two rectangles and look at the horizontal dimensions of each (Figure A.19). It follows that $a = 2$.

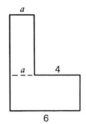

Figure A.19

Similarly, cut the figure vertically into two rectangles and focus on the vertical dimensions to see that $b + 5 = 8$, so $b = 3$ (Figure A.20).

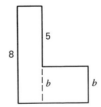

Figure A.20

The perimeter of the figure is $8 + 2 + 5 + 4 + 3 + 6 = 28$.

To find the perimeters of figures such as those in Exercises A-1 (page 228), you need to be able to compute the lengths of all the sides of the figures. In Chapter 10 of this book we look at the Pythagorean theorem and learn a method for computing the sides. You may wish to return to this section to practice using the Pythagorean theorem in the computation of perimeters.

Exercises A-2

I. Find the perimeter of each of the following polygons.

1.
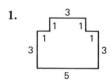

2.
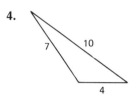

3.
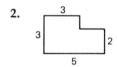

4.

5. Must figures with the same perimeter have the same area?

II. Find the perimeters of each of the following polygons.

1.
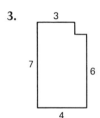

2.

3.
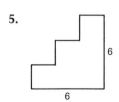

4.
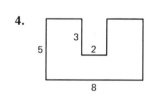

5.

6.
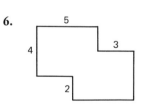

III. Solve the following problems.

1. **(a)** Use graph paper to show a rectangle with an area of 12 and a perimeter of 14.
   (b) Use graph paper to show a rectangle with an area of 12 and a perimeter of 16.

(c) Must rectangles with the same area have the same perimeter?

2. A rectangle has an area of 24 and a base of 8. Find its perimeter.

3. A rectangle has a perimeter of 24 and a base of 8. Find its area.

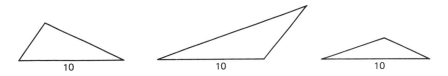

4. There are many triangles with base 10. None of them has a perimeter of 20 or less. Can you explain why?

APPENDIX B

CALCULATORS AND HIERARCHY OF OPERATIONS

PRETEST

Use a calculator to evaluate the following expressions.

1. $(17 + 6) \times 9$

2. $\dfrac{18 + 12}{3}$

3. $5^3 - 4^2$

4. $\dfrac{170}{3^4 + 4}$

Write a mathematical expression evaluated by each keying sequence.

5. $\boxed{4}$ $\boxed{+}$ $\boxed{6}$ $\boxed{\div}$ $\boxed{2}$ $\boxed{=}$

6. $\boxed{9}$ $\boxed{-}$ $\boxed{3}$ $\boxed{=}$ $\boxed{\times}$ $\boxed{5}$ $\boxed{=}$

7. $\boxed{8}$ $\boxed{y^x}$ $\boxed{3}$ $\boxed{-}$ $\boxed{1}$ $\boxed{=}$

8. $\boxed{6}$ $\boxed{+}$ $\boxed{4}$ $\boxed{x^2}$ $\boxed{=}$

9. In a keying sequence involving addition, multiplication, and squaring, which of the operations, in the absence of parentheses, does a calculator with hierarchy perform first?

ANSWERS TO PRETEST

1. Keying sequence: $\boxed{(}$ 17 $\boxed{+}$ $\boxed{6}$ $\boxed{)}$ $\boxed{\times}$ $\boxed{9}$ $\boxed{=}$

Or: 17 $\boxed{+}$ $\boxed{6}$ $\boxed{=}$ $\boxed{\times}$ $\boxed{9}$ $\boxed{=}$

Answer: 207

2. Keying sequence: 18 $\boxed{+}$ 12 $\boxed{=}$ $\boxed{\div}$ $\boxed{3}$ $\boxed{=}$

Or: $\boxed{(}$ 18 $\boxed{+}$ 12 $\boxed{)}$ $\boxed{\div}$ $\boxed{3}$ $\boxed{=}$

Answer: 10

3. Keying sequence: $\boxed{5}$ $\boxed{y^x}$ $\boxed{3}$ $\boxed{-}$ $\boxed{4}$ $\boxed{x^2}$ $\boxed{=}$

Answer: 109

4. Keying sequence: 170 $\boxed{\div}$ $\boxed{(}$ $\boxed{3}$ $\boxed{y^x}$ $\boxed{4}$ $\boxed{+}$ $\boxed{4}$ $\boxed{)}$ $\boxed{=}$

Or: $\boxed{3}$ $\boxed{y^x}$ $\boxed{4}$ $\boxed{+}$ $\boxed{4}$ $\boxed{=}$ $\boxed{STO}$ 170 $\boxed{\div}$ $\boxed{RCL}$ $\boxed{=}$

Answer: 2

5. $4 + \dfrac{6}{2}$ **6.** $(9 - 3) \times 5$

7. $8^3 - 1$ **8.** $6 + 4^2$

9. Squaring

If you missed:	*You need to study:*
1, 2, 5, or 6	Addition, subtraction, multiplication, and division (pages . . . to . . .).
3, 4, 7, or 8	Exponents on the calculator (pages . . . to . . .).
9	Better read through all of Appendix B.

B.1 ADDITION, SUBTRACTION, MULTIPLICATION, AND DIVISION

Using calculators we are often able to do numerical computations more quickly and accurately than we could do with pencil and paper alone. However, it is essential to understand how your calculator operates so that you will know how to direct it to do what you want it to do. The best reference on your calculator is the manual provided by the manufacturer. Here we point out some characteristics of calculators that should help you to get started.

Unfortunately, two different calculators may do different things with the same input. For example, if you key $\boxed{3}$ $\boxed{\times}$ $\boxed{4}$ $\boxed{+}$ $\boxed{5}$ $\boxed{\times}$ $\boxed{6}$ $\boxed{=}$, some calculators will display 102 and others will display 42. A calculator that displays 102 computes the operations in the order they are keyed: first it computes 3×4 to get 12, then it adds 5 to 12 to get 17, and finally it multiplies 17 by 6 to give 102. A calculator that displays 42 is using algebraic logic with **hierarchy.** On the second calculator, multiplication has priority over addition; going from left to right, the calculator computes all pending multiplications and divisions when either the $\boxed{+}$ or $\boxed{-}$ key is pressed. In our example, the calculator first computes 3×4, then 5 $\times$ 6, and finally computes the sum $12 + 30$ to get 42.

We assume in this book that you are using a calculator with hierarchy. If you are not, you need to interpret some of our statements for your calculator. Hierarchy on the calculator corresponds to the usual order of operations for arithmetic. It means that the expression $2 + 3 \times 4$ has only one meaning: $2 + 12$, or 14. If we want an expression that means "first add 2 and 3, and then multiply the sum by 4," we use parentheses and write $(2 + 3) \times 4$.

EXAMPLE 1   Use a calculator to evaluate $(7 + 9) \times 6$.

There are at least two ways to key this expression.

METHOD 1   The first is to use parentheses and key exactly what is written.

Keying sequence: $\boxed{(}\ \boxed{7}\ \boxed{+}\ \boxed{9}\ \boxed{)}\ \boxed{\times}\ \boxed{6}\ \boxed{=}$

Display: 0 · 7   7   9   16 16   6   96

Notice that pressing $\boxed{)}$ causes the calculator to perform the addition inside the parentheses; pressing $\boxed{=}$ causes the calculator to perform the multiplication.

METHOD 2   A second way to key the expression $(7 + 9) \times 6$ is to use the $\boxed{=}$ key to cause the addition to be performed before the multiplication.

Keying sequence: $\boxed{7}\ \boxed{+}\ \boxed{9}\ \boxed{=}\ \boxed{\times}\ \boxed{6}\ \boxed{=}$

Display: 7   7   9   16 16   6   96

EXAMPLE 2   Use a calculator to evaluate $\dfrac{2 + 6}{4}$.

You will see in Appendix D (or perhaps you remember from earlier study) that the bar in a fraction is a division symbol. The expression $\dfrac{2 + 6}{4}$ means that the sum of 2 and 6 is to be divided by 4.

METHOD 1   We can rewrite $\dfrac{2 + 6}{4}$ as $(2 + 6) \div 4$ and key this sequence.

Keying sequence: $\boxed{(}\ \boxed{2}\ \boxed{+}\ \boxed{6}\ \boxed{)}\ \boxed{\div}\ \boxed{4}\ \boxed{=}$

Display: 0   2   2   6   8   8   4   2

METHOD 2   We can use the $\boxed{=}$ key to perform the addition first in the expression $\dfrac{2 + 6}{4}$.

Keying sequence: $\boxed{2}\ \boxed{+}\ \boxed{6}\ \boxed{=}\ \boxed{\div}\ \boxed{4}\ \boxed{=}$

Display: 2   2   6   8   8   4   2

EXAMPLE 3   Use a calculator to compute $\dfrac{9 + 7}{5 + 3}$.

METHOD 1 Both of the following sequences are correct.

$$\boxed{[(}\ \boxed{9}\ \boxed{+}\ \boxed{7}\ \boxed{)]}\ \boxed{\div}\ \boxed{[(}\ \boxed{5}\ \boxed{+}\ \boxed{3}\ \boxed{)]}\ \boxed{=}$$

$$\boxed{9}\ \boxed{+}\ \boxed{7}\ \boxed{=}\ \boxed{\div}\ \boxed{[(}\ \boxed{5}\ \boxed{+}\ \boxed{3}\ \boxed{)]}\ \boxed{=}$$

If we were to omit all parentheses and key $\boxed{9}\boxed{+}\boxed{7}\boxed{\div}$ $\boxed{5}\boxed{+}\boxed{3}\boxed{=}$, the calculator would compute $9\ +\ 7\ \div$ $5 + 3$, that is, $9 + \dfrac{7}{5} + 3$.

METHOD 2 If your calculator has a memory, this is a good time to use it. The idea is to compute the denominator first, store it, do the rest of the computation, and recall the value of the denominator when you are ready to divide by it. Here is a possible keying sequence for $\dfrac{9 + 7}{5 + 3}$.

Keying
sequence: $\quad\boxed{5}\ \boxed{+}\ \boxed{3}\ \boxed{=}\ \boxed{\text{STO}}\ \boxed{9}\ \boxed{+}\ \boxed{7}\ \boxed{=}\ \boxed{\div}\ \boxed{\text{RCL}}\ \boxed{=}$

Display: $\qquad$ 5 $\quad$ 5 $\quad$ 3 $\quad$ 8 $\qquad$ 8 $\quad$ 9 $\quad$ 9 $\quad$ 7 $\quad$ 16 $\quad$ 16 $\quad$ 8 $\quad$ 2

The effect of the sequence is first to compute the denominator $5 + 3$ and store 8 in memory, then compute the numerator $9 + 7$, and finally divide the numerator by the denominator.

EXAMPLE 4 Give a mathematical expression evaluated by the keying sequence

$$\boxed{4}\ \boxed{\times}\ \boxed{3}\ \boxed{-}\ \boxed{8}\ \boxed{\div}\ \boxed{2}\ \boxed{=}$$

The calculator performs the multiplication when $\boxed{-}$ is pressed. When $\boxed{=}$ is pressed, it performs first $\boxed{\div}$ and then $\boxed{-}$. The expression evaluated is $(4 \times 3) - (8 \div 2)$.

EXAMPLE 5 Give a mathematical expression evaluated by the keying sequence

$$\boxed{3}\ \boxed{\times}\ \boxed{2}\ \boxed{\div}\ \boxed{6}\ \boxed{\times}\ \boxed{7}\ \boxed{=}$$

When the calculator receives a sequence of multiplication and division operations, it performs them in the order received. Thus, in this sequence the calculator computes 3 times 2 and divides the product by 6; that quotient is multiplied by 7.

$$\frac{3 \times 2}{6} \times 7$$

COMPUTING WITH NEGATIVE NUMBERS ON A CALCULATOR

Although we use the same symbol to denote negative numbers as we do for subtraction, most calculators have two keys: $\boxed{-}$ for subtraction and $\boxed{+/-}$ to change the sign of a number. The sign-change key $\boxed{+/-}$ changes the sign of any number on the display. To key the number -5, for example, first key the number 5 and then the sign-change key: $\boxed{5}$ $\boxed{+/-}$.

The examples in the following table illustrate how computations involving negative numbers can be keyed on a calculator.

Mathematical Expression	Keying Sequence	Display
$-5 + (-8)$	$\boxed{5}$ $\boxed{+/-}$ $\boxed{+}$ $\boxed{8}$ $\boxed{+/-}$ $\boxed{=}$	-13
$7 - (-4)$	$\boxed{7}$ $\boxed{-}$ $\boxed{4}$ $\boxed{+/-}$ $\boxed{=}$	11
$(-6) \times (-8)$	$\boxed{6}$ $\boxed{+/-}$ $\boxed{\times}$ $\boxed{8}$ $\boxed{+/-}$ $\boxed{=}$	48
$9 \div (-3)$	$\boxed{9}$ $\boxed{\div}$ $\boxed{3}$ $\boxed{+/-}$ $\boxed{=}$	-3

Exercises B-1

I. Use your calculator to evaluate each of the following expressions.

1. $2 + 5 \times 6$

2. $(2 + 5) \times 6$

3. $5 \times (1 + 2 + 3)$

4. $12 + 4 \times (3 + 6)$

5. $7 + 3 \div 3 + 2$

6. $(7 + 3) \div (3 + 2)$

7. $7 \times 2 - 4 \times 3$

8. $\dfrac{18}{2 + 4}$

9. $\dfrac{16 + 8}{4}$

10. $\dfrac{192 - 12}{3 + 12}$

11. $15 - (-10)$

12. $(-138) \div (-6)$

II. Write a mathematical expression evaluated by each of the following keying sequences.

1. $\boxed{4}$ $\boxed{+}$ $\boxed{6}$ $\boxed{\div}$ $\boxed{3}$ $\boxed{=}$

2. $\boxed{8}$ $\boxed{\times}$ $\boxed{7}$ $\boxed{-}$ $\boxed{4}$ $\boxed{\times}$ $\boxed{2}$ $\boxed{=}$

3. $\boxed{8}$ $\boxed{\div}$ $\boxed{2}$ $\boxed{\times}$ $\boxed{2}$ $\boxed{=}$

4. $\boxed{9}$ $\boxed{+}$ $\boxed{6}$ $\boxed{=}$ $\boxed{\div}$ $\boxed{3}$ $\boxed{=}$

5. $\boxed{8}$ $\boxed{\div}$ $\boxed{(}$ $\boxed{3}$ $\boxed{+}$ $\boxed{1}$ $\boxed{)}$ $\boxed{=}$

6. $\boxed{7}$ $\boxed{-}$ $\boxed{5}$ $\boxed{=}$ $\boxed{STO}$ $\boxed{9}$ $\boxed{+}$ $\boxed{5}$ $\boxed{=}$ $\boxed{\div}$ $\boxed{RCL}$ $\boxed{=}$

7. $\boxed{(}$ $\boxed{4}$ $\boxed{+}$ $\boxed{7}$ $\boxed{)}$ $\boxed{\times}$ $\boxed{6}$ $\boxed{=}$

8. $\boxed{8}$ $\boxed{-}$ $\boxed{2}$ $\boxed{\times}$ $\boxed{3}$ $\boxed{=}$

9. $\boxed{5}$ $\boxed{+/-}$ $\boxed{+}$ $\boxed{9}$ $\boxed{=}$

B.2 EXPONENTS ON THE CALCULATOR

A number times itself is often read as the number squared and written using the exponent 2; that is, $5 \times 5 = 5^2$ is read as "five squared." A calculator often has a special key $\boxed{x^2}$ to compute the square of a number. To compute 5^2, for example, key $\boxed{5}\boxed{x^2}$. The value of 5^2 is immediately displayed; the $\boxed{=}$ key need not be used.

Larger exponents are used when a number appears in a product more than twice. For example,

$$5^3 = 5 \times 5 \times 5$$

$$5^4 = 5 \times 5 \times 5 \times 5$$

(See Chapter 2 for a complete discussion of exponents.) A scientific calculator has a key $\boxed{y^x}$ to provide for any exponent. For example, to evaluate 5^3, key in

$$\boxed{5}\ \boxed{y^x}\ \boxed{3}\ \boxed{=}$$

To evaluate 5^4, key in

$$\boxed{5}\ \boxed{y^x}\ \boxed{4}\ \boxed{=}$$

In the calculator's hierarchy (and in the arithmetic order of operations) the $\boxed{x^2}$ and the $\boxed{y^x}$ keys take precedence over multiplication and division. In the absence of parentheses, raising a number to a power is performed before multiplication and division; multiplication and division, as you know, are performed before addition and subtraction.

EXAMPLE 1   Use a calculator to evaluate $4^3 - 3^2$.

The sequence can be keyed as written. Watch your calculator display to see that raising to a power is performed in both cases before the subtracting.

Keying sequence: $\boxed{4}\ \boxed{y^x}\ \boxed{3}\ \boxed{-}\ \boxed{3}\ \boxed{x^2}\ \boxed{=}$

Display: 4   4   3   64   3   9   55

EXAMPLE 2   Use a calculator to evaluate $\dfrac{7^2 - 1}{2^3}$.

Remember that the whole expression $7^2 - 1$ is divided by 2^3.

Keying sequence: $\boxed{7}\ \boxed{x^2}\ \boxed{-}\ \boxed{1}\ \boxed{=}\ \boxed{\div}\ \boxed{2}\ \boxed{y^x}\ \boxed{3}\ \boxed{=}$

Display: 7   49   49   1   48   48   2   2   3   6

Another way of doing this computation on your calculator is

Keying sequence: $\boxed{(}\ \boxed{7}\ \boxed{x^2}\ \boxed{-}\ \boxed{1}\ \boxed{)}\ \boxed{\div}\ \boxed{2}\ \boxed{y^x}\ \boxed{3}\ \boxed{=}$

Display: 0   7   49   49   1   48   48   2   2   3   6

EXAMPLE 3   Write a mathematical expression evaluated by the following sequence.

$$\boxed{5}\ \boxed{+}\ \boxed{3}\ \boxed{y^x}\ \boxed{4}\ \boxed{=}$$

The calculator performs the operation of raising to a power before the addition. Only 3 is raised to the power 4. Thus, the expression $5 + 3^4$ is evaluated.

EXAMPLE 4   Write a mathematical expression evaluated by this sequence.

$$\boxed{5}\ \boxed{+}\ \boxed{3}\ \boxed{=}\ \boxed{y^x}\ \boxed{4}\ \boxed{=}$$

If you watch the display on your calculator, you will see that the addition is performed when the first $\boxed{=}$ is pressed. Thus, the keying sequence evaluates $(5 + 3)^4$. Another way to evaluate this expression is to key

$$\boxed{(}\ \boxed{5}\ \boxed{+}\ \boxed{3}\ \boxed{)}\ \boxed{y^x}\ \boxed{4}\ \boxed{=}$$

Exercises B-2

I. Use a calculator to evaluate these expressions.

1. $2 \times 3^2 + 5^2$

2. $4^5 - 5^4$

3. $\dfrac{7^4 + 2^3}{3}$

4. $6^5 \times 3^2$

5. $(12 + 8)^2$

6. $\left(\dfrac{21 - 3}{6}\right)^3$

7. $3 \times (7 + 4)^2 - 6^3$

8. $(5 \times 2)^2 + (4 \times 3)^2$

II. Write a mathematical phrase evaluated by each of the following keying sequences.

1. $\boxed{7}$ $\boxed{x^2}$ $\boxed{-}$ $\boxed{2}$ $\boxed{=}$

2. $\boxed{5}$ $\boxed{y^x}$ $\boxed{4}$ $\boxed{-}$ $\boxed{1}$ $\boxed{=}$

3. $\boxed{4}$ $\boxed{y^x}$ $\boxed{(}$ $\boxed{6}$ $\boxed{-}$ $\boxed{4}$ $\boxed{)}$ $\boxed{=}$

4. $\boxed{3}$ $\boxed{^+/_-}$ $\boxed{x^2}$ $\boxed{-}$ $\boxed{2}$ $\boxed{x^2}$ $\boxed{=}$

5. $\boxed{9}$ $\boxed{+}$ $\boxed{3}$ $\boxed{x^2}$ $\boxed{=}$

6. $\boxed{9}$ $\boxed{+}$ $\boxed{3}$ $\boxed{=}$ $\boxed{x^2}$ $\boxed{^+/_-}$

7. $\boxed{3}$ $\boxed{x^2}$ $\boxed{x^2}$

8. $\boxed{6}$ $\boxed{x^2}$ $\boxed{\div}$ $\boxed{4}$ $\boxed{=}$ $\boxed{y^x}$ $\boxed{3}$ $\boxed{=}$

APPENDIX C

PERCENT

PRETEST

1. Find 19% of 24,356.
2. What percent of 540 is 27?
3. 18 is what percent of 90?
4. A "stop smoking" clinic advertizes that 67% of those completing its course are still nonsmokers after one year. A group of 60 people have just finished the course; how many of these people will probably be nonsmokers a year from now?
5. A store raised the price of an item from $36 to $45. By what percent was the price increased?

ANSWERS TO PRETEST

1. 4627.64
2. 5%
3. 20%
4. 40
5. 25%

If you answered 1, 2, and 3 correctly but missed 4 or 5, you need to review Section C.3.

If you missed 3 or more problems, you should review the entire appendix.

C.1   INTRODUCTION TO PERCENT

The word **percent** means per hundred; thus, the phrase 40% means 40 per hundred, 51% means 51 per hundred, 123% means 123 per hundred, and $\frac{1}{2}$% means $\frac{1}{2}$ per hundred. The meaning of these phrases becomes clearer when we examine them in applied contexts; consider the following examples.

EXAMPLE 1   The A to Z Market is having its giant sale—40% off on every item in the store!

Here the 40% refers to the regular price of each item. For every hundred cents in the original price, 40 cents is subtracted to determine the sale price. Thus, for an item which regularly costs $5, the sale price will be determined by subtracting 40¢ for each dollar; that is, $5 − 5(0.40) = $3.

EXAMPLE 2   51% of the babies born in 1979 were girls.

Here percent refers to the number of babies born in 1979; for every hundred born, 51 were girls (and 49 were boys). Suppose 35,000 babies were born in February 1979; of these, the number that were girls can be computed as follows. There were $\frac{35,000}{100} = 350$ hundreds of babies; since 51 per hundred were girls, 51 · 350 or 17,850 were girls.

EXAMPLE 3   The cost of steak this year is 123% of last year's cost.

Since $1 is 100 cents, for every dollar the steak cost last year, it would cost 123 cents or $1.23 this year. Thus, a steak that cost $4 last year would cost 4($1.23) or $4.92 this year.

EXAMPLE 4   An income tax of $\frac{1}{2}$% goes to support the Public Library.

Here the percent refers to a person's income. For every hundred dollars of income, $\frac{1}{2}$ dollar is paid for this tax. A person earning $15,000 would pay 150 half-dollars or $75 tax toward the library.

Notice that in order to apply percent to a situation you must first determine to what the percent refers; this referent is often called the **base.** If the base is B, then A% of B is equal to $\frac{A}{100} \cdot B$. The division by 100 reflects the fact that A% means A per hundred. Dividing a number by 100 has the effect of "moving the decimal point" two places to the left. Suppose we wanted to calculate 43% of 58; 43% of 58 equals $\frac{43}{100} \cdot 58$. Since $\frac{43}{100} = 0.43$, the product can also be written as $0.43 \cdot 58$. The latter method is more practical when you are using a calculator.

Exercises C-1

I. Compute the following.

   1. 75% of 32   **2.** 23% of 48

3. 29% of 15

4. 116% of 82

5. 53% of 123

6. $\frac{1}{2}$% of 240

7. $\frac{1}{10}$% of 13

8. 20% of 0.02

9. $2\frac{1}{2}$% of 28

10. $11\frac{3}{4}$% of 12,000

II. Describe the base of the percent for each of the following phrases.

1. John got a 15% raise.

2. The interest rate for passbook savings accounts is 6%.

3. The interest rate for automobile loans is $13\frac{1}{2}$%.

4. The property tax is 1.5%.

5. Voter registration is up 20% this year.

6. From 1970 to 1980, membership increased by 13%.

7. The return rate for library books is 87%.

8. A fruit drink is 8% juice.

9. The city wage tax is 2%.

10. Ivory soap is $99\frac{44}{100}$% pure.

C.2   USING PERCENT

The typical percent situation involves a certain percent of one number (the base) that is equal to another number; written symbolically,

$$A\% \text{ of } B \text{ is } C$$

or

$$\frac{A}{100} \cdot B = C$$

where A, B, and C are numbers. There are three questions that can be asked in this situation.

1. What percent of B is C?   (Answer: $A\%$)
2. What is $A\%$ of B?   (Answer: C)
3. C is $A\%$ of what number?   (Answer: B)

If any two of the numbers A, B, and C are known, the third can be calculated. Study the following examples carefully.

EXAMPLE 1   What is 35% of 420?

$$35\% \text{ of } 420 \text{ is } C$$

$$\frac{35}{100} \cdot 420 = C$$

$$0.35 \cdot 420 = C$$

$$147 = C$$

EXAMPLE 2   15% of what number is 48.75?

$$15\% \text{ of } B \text{ is } 48.75$$

$$\frac{15}{100} \cdot B = 48.75$$

$$0.15 \cdot B = 48.75$$

$$B = \frac{48.75}{0.15}$$

$$B = 325$$

.15x = 48.75
x = 48.75 / .15
x = 325

EXAMPLE 3   What percent of 72 is 27?

$$A\% \text{ of } 72 \text{ is } 27$$

$$\frac{A}{100} \cdot 72 = 27$$

$$\frac{72}{100} \cdot A = 27$$

$$0.72 \cdot A = 27$$

$$A = \frac{27}{0.72}$$

$$A = 37.5\%$$

Each application of percent involves one of these forms. Sometimes other arithmetic must also be done to solve a problem, or you may have to use percent more than once, but the percent parts of the problem always take one of these forms.

Exercises C-2

Compute answers to the following questions.
1. What is 5% of 130?   *6.5*
2. 17 is 10% of what number?   *.10 x = 17*
   x = 170

3. What is 43% of 18?   *7.74*
4. 12% of 18 is what number?   *2.16*
5. What percent of 24 is 18?   *75%*
6. What number is 43% of 92?   *39.56*
7. 24 is 25% of what number?   *96*
8. What number is 0.05% of 83?   *4.15*
9. 86 is what percent of 25?   *344%*
10. 53 is what percent of 106?   *50%*
11. 18% of what number is 36?   *.18x = 36   200*
12. $\frac{1}{2}$% of what number is 23?   *4600*
13. 154% of what number is 100?   *64.9351*
14. 28 is what percent of 78?   *36%*
15. What is 4% of 15% of 80?   *.48*
16. When a certain number was increased by 25%, the result was 85. What was the original number?   *68*
17. When a certain number was decreased by 15%, the result was 170. What was the original number?   *200*
18. What is 105% of 105% of 180?   *198.45*
19. What is 5% of 5% of 180?   *.45*
20. 27 is 80% of 60% of what number?   *27 = .48x   x = 56.25*

C.3 APPLICATIONS OF PERCENT

Many times problems involving percent seem to be difficult. What makes them hard is not so much the concept of percent as it is remembering how this concept is used in applications. In order to do problems involving percent, you should begin by analyzing the problem to find the base. Consider the following example.

A department store has just marked down all the items in its linen department by 22%. What is the sale price of a pillow that originally cost $19?

To do this problem one must know that "marked down by 22%" means that the price has been reduced by 22 percent of the original price. Therefore, the effect of the markdown on the price of the pillow mentioned is that 22% of $19 is subtracted from the original price of $19.

$$\text{New price} = 19 - 0.22 \cdot 19$$
$$= 19 - 4.18$$
$$= 14.82$$

Thus, the effect of the markdown is to reduce the price of the pillow to $14.82.

Many applications of percent involve a percent change from one figure (the original) to another. In general, these percents are determined by using the original figure as a base. Consider, for example, the following problem.

Jerre just received a 12% raise in salary. Her new salary is $14,000. What was her salary before the raise?

In order to do this problem you must first realize that Jerre's raise (and all such raises) was computed on the basis of her old salary. Suppose Jerre's old salary was x dollars. The accountant figuring her new salary would first find 12% of x dollars ($0.12x$) and then add it to the old salary, arriving at the figure $x + 0.12x$ as her new salary. The problem asserts that her new salary is $14,000. Therefore,

$$x + 0.12x = 14,000$$
$$(1 + 0.12)x = 14,000 \qquad \text{(Distributive principle)}$$
$$1.12x = 14,000 \qquad \text{(Addition)}$$
$$x = \frac{14,000}{1.12} \qquad \text{(Divide both sides by 1.12)}$$
$$x = \$12,500 \qquad \text{(Computation)}$$

Thus, Jerre's old salary was $12,500.

It is important to note that the problem *cannot* be solved by finding 12% of her new salary and subtracting. 12% of 14,000 is 1680; $14,000 - 1680 = 12,320$, which is *not* the same as $12,500. This difference illustrates the importance of using the appropriate base.

Exercise C-3

1. The hourly rate for babysitters is 75¢ before midnight. In a certain community it is customary to increase this by 50% after midnight. What is the hourly rate after midnight?

2. A fruit drink is 11% juice; the rest is water. How much juice is required to make 500 gallons of the drink.

3. The Majors are in the "35% tax bracket." How much tax must they pay on $42,000 of taxable income.

4. Joan Terry receives $1024 per month in take-home pay. Her monthly rent is $320. What percent of her take-home pay goes for rent.

5. The property tax on a $50,000 home in Crestville is $1800 per year. What is the tax rate on residences.

6. Max Beck was earning $18,000 until last January when he was given a 16% raise. What was his salary after the raise?

7. Frankie was just promoted and given a 16% raise. His new salary is $16,820. What was his old salary?

8. The Localsville police force has recently been cut by 36% in a money-saving move. The current force is 128 officers. How large was the force before the cuts?

9. Mary received a 10% raise in 1979 and an 18% raise in 1980. By what percent was her salary raised in the 2-year period?

10. Ms. Smith wanted to return to work. To facilitate this move, in August her husband gave up his overtime work decreasing the family income by 17%. In October, Ms. Smith got a job that increased their September income by 79%. Overall what effect did these changes have on the family income?

11. The bill in a fancy restaurant was $45.80 for two business men. The first man added a 15% tip to the bill and charged the entire cost. The second man saw the total amount charged and, not realizing that it included a tip, left a tip equal to 15% of the grand total. What percent tip did the lucky waiter receive?

APPENDIX D

FRACTIONS

PRETEST

Compute the following.

1. $\dfrac{1}{2} + \dfrac{1}{3}$

2. $\dfrac{5}{12} + \dfrac{3}{8}$

3. $1\dfrac{2}{3} + 1\dfrac{1}{6}$

4. $\dfrac{2}{5} - \dfrac{1}{7}$

5. $\dfrac{3}{4} \cdot \dfrac{5}{8}$

6. $1\dfrac{3}{8} \cdot \dfrac{2}{3}$

7. $\dfrac{7}{8} \div \dfrac{1}{4}$

8. $\dfrac{5}{12} \div \dfrac{4}{5}$

9. Indicate how to key $\dfrac{5}{8}$ on a calculator.

10. Arrange these fractions in order, from largest to smallest: $\dfrac{4}{3}, \dfrac{12}{13}, \dfrac{1}{2}, \dfrac{1}{3}, \dfrac{12}{11},$ and $\dfrac{2}{3}$.

ANSWERS TO PRETEST

1. $\dfrac{5}{6}$

2. $\dfrac{19}{24}$ or $\dfrac{76}{96}$

3. $2\dfrac{5}{6}$ or $\dfrac{17}{6}$ or $2\dfrac{15}{18}$ or $\dfrac{51}{18}$

4. $\dfrac{9}{35}$

5. $\dfrac{15}{32}$

6. $\dfrac{22}{24}$ or $\dfrac{11}{12}$

7. $\dfrac{28}{8}$ or $\dfrac{14}{4}$ or $\dfrac{7}{2}$

8. $\dfrac{25}{48}$

9. $\boxed{5}\,\boxed{\div}\,\boxed{8}\,\boxed{=}$

10. $\dfrac{4}{3}, \dfrac{12}{11}, \dfrac{12}{13}, \dfrac{2}{3}, \dfrac{1}{2}, \dfrac{1}{3}$

If you missed:	You need to study:
1, 2, 3, 4	Addition and subtraction (pages 000 to 000).
5, 6, 7, 8	Multiplication and division (pages 000 to 000).
3, 6	Fractions greater than 1 (pages 000 to 000).
9	Meaning of fractions (pages 000 to 000).
10	Ordering of fractions (pages 000 to 000).
More than 4 problems	Better study the entire appendix.

D.1   REPRESENTING FRACTIONS

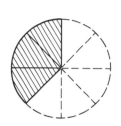

Figure D.1

There are many ways of thinking about the meaning of a fraction such as $\frac{3}{8}$. One of the first ways children are taught about fractions, and a way that many adults remember most easily, is in terms of parts of a pie or cake.

Look at Figure D.1. $\frac{3}{8}$ of a cake is the amount of cake left if the cake is cut into 8 pieces of equal size and all but 3 of the pieces are eaten.

In applying fractions to problems and to the development of more complex mathematics, two other interpretations are more useful. The first of these involves considering the fraction as representing a part of a square region.

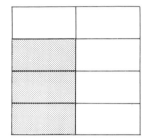

Figure D.2

Now take a look at Figure D.2. $\frac{3}{8}$ of a square region is the part of the region that is shaded if the region is divided into 8 identical parts and 3 of the parts are shaded.

We will use this representation in parts of this appendix; when we do, we will assume the square has area 1.

Figure D.3

A second useful representation of the fraction $\frac{3}{8}$ is as a point on the number line. In the first four chapters of this book, the integers were identified with points on a number line, as shown in Figure D.3. To place the fraction $\frac{3}{8}$ on this number line, observe first that $\frac{3}{8}$ is greater than 0 and that it is less than 1; therefore, its position on the line should be between 0 and 1. The exact position can be determined by analogy with the cake and square representations above.

$\frac{3}{8}$ is the point on the number line determined by first dividing the interval from 0 to 1 into 8 identical pieces. See Figure D.4.

Figure D.4

Then, beginning at 0, count off 3 of these pieces, as in Figure D.5.

Figure D.5

$\frac{3}{8}$ is the point at the end of the three pieces when we count from 0, as illustrated in Figure D.6.

Figure D.6

Another important interpretation of the fraction $\frac{3}{8}$ is as the quotient $3 \div 8$. It is hard to draw a diagram showing this interpretation without recalling the relationship between multiplication and division. Remembering that multiplication "undoes" division so that $(3 \div 8) \cdot 8 = 3$, consider the following situation illustrating $\left(\frac{3}{8}\right) \cdot 8$.

Suppose we took the remains of 8 cakes, each with $\frac{3}{8}$ left, and rearranged the pieces to form whole cakes. We would find that we had 3 whole cakes. Examine Figure D.7.

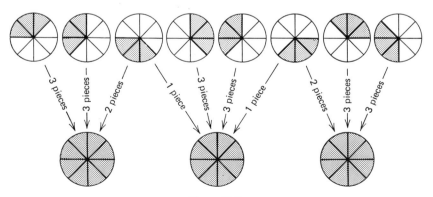

Figure D.7

Thus, $\frac{3}{8} \cdot 8 = 3$.

Since $\frac{3}{8} \cdot 8 = 3$ and $(3 \div 8) \cdot 8 = 3$, we conclude that $\frac{3}{8} = 3 \div 8$.

If a and b are whole numbers with a less than b, we can consider the fraction $\frac{a}{b}$ as having meanings corresponding to those for $\frac{3}{8}$. For example, consider another fraction, $\frac{2}{5}$.

$\frac{2}{5}$ is the amount of cake left if the cake is divided into 5 pieces of equal size and all but 2 of them are eaten.

$\frac{2}{5}$ of a square region is the part of the region that is shaded if the region is divided into 5 identical parts and 2 of the parts are shaded.

$\frac{2}{5}$ is the number that corresponds to the point between 0 and 1 on the number line; it can be found by dividing the unit interval into 5 identical parts and finding the right-hand end of the second part.

$\frac{2}{5}$ is the quotient $2 \div 5$.

EQUIVALENT FRACTIONS

If we carefully examine fractions and their meanings, we find that some pairs of fractions have the same meaning. We say that such fractions are **equivalent.** Consider the following examples.

EXAMPLE 1 $\frac{1}{4}$ cake is left $\frac{2}{8}$ cake is left

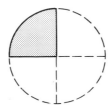   

$\frac{1}{4}$ and $\frac{2}{8}$ describe the same amount of cake, so $\frac{1}{4}$ and $\frac{2}{8}$ are equivalent; that is, $\frac{1}{4} = \frac{2}{8}$. $\frac{1}{4} = \frac{2}{8}$ and $\frac{2}{8} = \frac{2 \cdot 1}{2 \cdot 4}$.

EXAMPLE 2 $\frac{2}{3}$ of the square is shaded $\frac{6}{9}$ of the square is shaded

 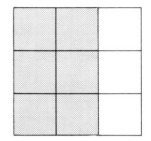

$\frac{2}{3}$ and $\frac{6}{9}$ describe the same part of the square region, so $\frac{2}{3}$ is equivalent to $\frac{6}{9}$; that is, $\frac{2}{3} = \frac{6}{9}$. $\frac{2}{3} = \frac{6}{9}$ and $\frac{6}{9} = \frac{3 \cdot 2}{3 \cdot 3}$.

EXAMPLE 3

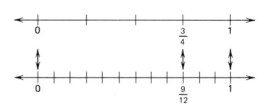

$\frac{9}{12}$ and $\frac{3}{4}$ identify the same point on the number line, so $\frac{9}{12}$ is equivalent to $\frac{3}{4}$, or $\frac{9}{12} = \frac{3}{4}$. $\frac{3}{4} = \frac{9}{12}$ and $\frac{9}{12} = \frac{3 \cdot 3}{3 \cdot 4}$.

EXAMPLE 4

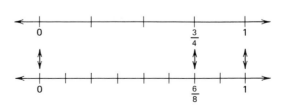

$\frac{6}{8}$ and $\frac{3}{4}$ identify the same point on the number line, so $\frac{6}{8}$ is equivalent to $\frac{3}{4}$, or $\frac{6}{8} = \frac{3}{4}$. $\qquad \frac{6}{8} = \frac{3}{4}$ and $\frac{6}{8} = \frac{2 \cdot 3}{2 \cdot 4}$.

These examples lead us to a general principle that describes equivalent fractions.

If $\frac{a}{b}$ is a fraction and $n \neq 0$, then $\frac{a}{b} = \frac{n \cdot a}{n \cdot b}$.

If we examine Examples 3 and 4 carefully, we see that $\frac{6}{8}$ must equal $\frac{9}{12}$ since each of these fractions is equal to $\frac{3}{4}$. The principle of equivalent fractions does not show this equivalence directly because 9 is not a multiple of 6. Often we will wish to know whether two fractions are equivalent. The procedure known as **cross-multiplication** provides a test. This procedure is essentially a short way of finding fractions with the same denominator (bottom number) equivalent to the two given fractions; numerators (top numbers) are then compared. See if you can follow this example for $\frac{6}{8}$ and $\frac{9}{12}$.

$$\frac{6}{8} = \frac{12 \cdot 6}{12 \cdot 8}$$

Principle of equivalent fractions with $n = 12$, the denominator of $\frac{9}{12}$.

$$\frac{9}{12} = \frac{8 \cdot 9}{8 \cdot 12}$$

Principle of equivalent fractions with $n = 8$, the denominator of $\frac{6}{8}$.

The fractions $\frac{12 \cdot 6}{12 \cdot 8}$ and $\frac{8 \cdot 9}{8 \cdot 12}$ have the same denominator. Also

$$12 \cdot 6 = 72$$
$$8 \cdot 9 = 72$$

so the fractions have the same numerator. Therefore, the fractions are equal and $\frac{6}{8} = \frac{9}{12}$.

Notice that we did not have to compute $12 \cdot 8$ and $8 \cdot 12$. The commutative principle guarantees that they are equal. We needed only to compute $12 \cdot 6$ and $8 \cdot 9$. A general rule for the cross-multiplication procedure follows.

If $\dfrac{a}{b}$ and $\dfrac{c}{d}$ are fractions, then $\dfrac{a}{b} = \dfrac{c}{d}$ provided $a \cdot d = b \cdot c$.

The following examples illustrate the use of this principle.

EXAMPLE 1   Compare $\dfrac{3}{15}$ with $\dfrac{5}{25}$.

In these fractions, we have $a = 3$, $b = 15$, $c = 5$, and $d = 25$. Then $a \cdot d = 3 \cdot 25 = 75$ and $b \cdot c = 15 \cdot 5 = 75$. Thus, $a \cdot d = b \cdot c$ and the fractions are equal.

EXAMPLE 2   Compare $\dfrac{8}{12}$ with $\dfrac{24}{32}$.

We let $a = 8$, $b = 12$, $c = 24$, and $d = 32$. Then $a \cdot d = 8 \cdot 32 = 256$ and $b \cdot c = 12 \cdot 24 = 288$. Thus, $a \cdot d \neq b \cdot c$ and $\dfrac{8}{12} \neq \dfrac{24}{32}$. These fractions are not equal.

The reason for the name **cross-multiplication** is related to the way some people remember what numbers to multiply together. Look at the next example.

EXAMPLE 3   Compare $\dfrac{9}{12}$ and $\dfrac{3}{20}$.

$$\frac{9}{12} \diagdown\!\!\!\!\!\diagup \frac{13}{20}$$

$$\left.\begin{array}{l} 9 \cdot 20 = 180 \\ 12 \cdot 13 = 156 \end{array}\right\} \quad \text{unequal}$$

Thus, the fractions are not equal.

REDUCING FRACTIONS

There is an important consequence to the statement that, if $n \neq 0$, a fraction $\dfrac{a}{b} = \dfrac{na}{nb}$. The fact $\dfrac{na}{nb} = \dfrac{a}{b}$ means that whenever the numerator and denominator of a fraction both contain the same factor n, that factor n can be removed from the numerator and denominator without changing the value of the fraction.

$$\frac{2 \cdot 7}{2 \cdot 15} = \frac{7}{15}$$

The process of removing common factors from the numerator and denominator is called **reducing the fraction.** If all common factors have been removed from the numerator and denominator, we say the fraction is in **reduced form.** To write a fraction in reduced form, we first factor the numerator and denominator into a product of prime numbers and then remove common factors.

EXAMPLE 4   Write the fraction $\frac{42}{90}$ in reduced form.

Since $42 = 6 \cdot 7 = 2 \cdot 3 \cdot 7$ and $90 = 9 \cdot 10 = 3 \cdot 3 \cdot 2 \cdot 5$, we have

$$\frac{42}{90} = \frac{2 \cdot 3 \cdot 7}{3 \cdot 3 \cdot 2 \cdot 5} = \frac{7}{3 \cdot 5} = \frac{7}{15}$$

The common factors 2 and 3 have been removed from the numerator and the denominator.

Exercises D-1

I. For each of the following fractions, describe its meaning using each of the four interpretations given on pages 255 and 256.

1. $\dfrac{5}{6}$

2. $\dfrac{1}{2}$

3. $\dfrac{7}{12}$

4. $\dfrac{7}{9}$

5. $\dfrac{3}{7}$

6. $\dfrac{6}{12}$

II. For each of the following fractions, find five others that are equivalent to it.

1. $\dfrac{1}{3}$

2. $\dfrac{3}{4}$

3. $\dfrac{5}{8}$

4. $\dfrac{69}{96}$

5. $\dfrac{540}{630}$

6. $\dfrac{968}{990}$

7. $\dfrac{3a}{2b}$

8. $\dfrac{a^2}{b^2}$

III. Express each quotient as a fraction.

1. $5 \div 40$ 4. $9 \div 15$
2. $5 \div 41$ 5. $8 \div 13$
3. $7 \div 6$

IV. Express each fraction as a quotient and compute its decimal value on your calculator.

1. $\dfrac{3}{5}$ 7. $\dfrac{42}{7}$

2. $\dfrac{6}{11}$ 8. $\dfrac{101}{11}$

3. $\dfrac{9}{5}$ 9. $5\dfrac{8}{15}$

4. $\dfrac{12}{25}$ 10. $7\dfrac{9}{25}$

5. $\dfrac{13}{19}$ 11. $3\dfrac{3}{10}$

6. $\dfrac{25}{19}$ 12. $25\dfrac{1}{25}$

V. Use the cross-multiplication procedure to decide if the fractions in each pair are equal.

1. $\dfrac{4}{6}$ $\dfrac{26}{39}$ 6. $\dfrac{6}{8}$ $\dfrac{75}{100}$

2. $\dfrac{40}{80}$ $\dfrac{12}{24}$ 7. $\dfrac{9}{12}$ $\dfrac{100}{125}$

3. $\dfrac{7}{8}$ $\dfrac{8}{9}$ 8. $\dfrac{16}{36}$ $\dfrac{64}{144}$

4. $\dfrac{15}{20}$ $\dfrac{45}{60}$ 9. $\dfrac{99}{100}$ $\dfrac{9}{10}$

5. $\dfrac{8}{30}$ $\dfrac{12}{45}$ 10. $\dfrac{88}{256}$ $\dfrac{77}{224}$

VI. Write each of the following fractions in reduced form.

1. $\dfrac{14}{26}$ 5. $\dfrac{71}{132}$

2. $\dfrac{8}{20}$ 6. $\dfrac{30}{75}$

3. $\dfrac{12}{15}$ 7. $\dfrac{64}{100}$

4. $\dfrac{35}{84}$ 8. $\dfrac{39}{42}$

VII. Give a fraction as the answer in each of the following questions.

1. Every day John runs 40 laps around the track. At the moment he has just finished his 24th lap. What part of today's run has he finished?

2. A test consisted of 50 true-or-false questions. Mary skipped 2 questions and gave wrong answers to 4 questions. What part of the test did Mary answer correctly?

3. Tom poured 1 quart of water and 1 cup of juice concentrate into an empty half-gallon jar. How full was his jar?

D.2   FRACTIONS GREATER THAN 1

The fractions we have considered thus far have all had numerators (top numbers) less than their denominators (bottom numbers); these fractions all identified points on the number line between 0 and 1. We now look further at our number line model to examine the meaning of fractions with numerators greater than or equal to their denominators. Consider the fractions $\frac{3}{4}$ and $\frac{7}{4}$ by dividing the unit interval between 0 and 1 into 4 identical parts. To locate $\frac{3}{4}$, we find the point at the right-hand end of the third part or segment, as in Figure D.8.

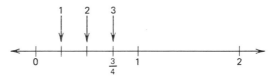

Figure D.8

We cannot find $\frac{7}{4}$ between 0 and 1; to find it we need to continue marking off identical segments beyond 1 until we have 7 of them, as in Figure D.9.

Figure D.9

Notice that we can place any fraction with a denominator of 4 and positive numerator on this number line (Figure D.10).

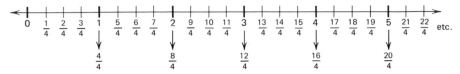

Figure D.10

By a similar procedure we could place any positive fraction on the number line. Figure D.11 illustrates some fractions with a denominator of 6.

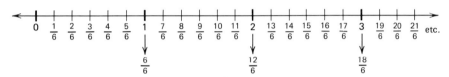

Figure D.11

You will notice on the number lines in Figures D.10 and D.11 that $\frac{6}{6} = 1$ and $\frac{8}{4} = 2$. Every whole number can be represented as a fraction with any denominator. For example, 7 is equal to $\frac{7}{1}, \frac{14}{2}, \frac{21}{3}, \frac{70}{10}, \frac{77}{11}, \frac{630}{90}$, and many other fractions.

The observation above leads us to another way of writing fractions greater than 1. Look again at our graphs of $\frac{7}{4}$ and $\frac{3}{4}$; the point associated with $\frac{7}{4}$ can be found by locating 1 and then marking the point that is $\frac{3}{4}$ unit to the right of 1. This, as we will see, is $1 + \frac{3}{4}$. Thus, $\frac{7}{4} = 1 + \frac{3}{4}$. See Figure D.12.

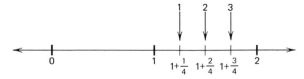

Figure D.12

In practice we usually write $\frac{7}{4} = 1\frac{3}{4}$, omitting the "+" sign from the expression. To write a fraction that is greater than 1 as a whole number plus a fraction between 0 and 1, we recall that a fraction is a quotient of whole numbers. Consider, for example, the fraction $\frac{29}{3}$, and observe that $\frac{29}{3} = \frac{27 + 2}{3}$. We can view this as $\frac{27}{3}$ plus $\frac{2}{3}$; $\frac{27}{3} = 9$, so $\frac{27}{3} + \frac{2}{3} = 9 + \frac{2}{3} = 9\frac{2}{3}$. Consider, now the fraction $\frac{380}{13}$; it is not as easy to find the whole number part of this fraction just by looking at it. If we divide 380 by 13, we obtain

$$
\begin{array}{r}
29 \\
13)\overline{380} \\
\underline{26} \\
120 \\
\underline{117} \\
3
\end{array}
$$

or $380 = 13 \cdot 29 + 3$. This means that $\frac{380}{13} = \frac{13 \cdot 29 + 3}{13} = \frac{13 \cdot 29}{13} + \frac{3}{13} = 29 + \frac{3}{13}$ or $29\frac{3}{13}$. A similar division procedure can be used to convert any fraction greater than 1 to a "mixed number."

To go the other way—that is, to write a number such as $7\frac{3}{8}$ as a single fraction—we reverse this procedure. To see why this works consider the examples.

$$7\frac{3}{8} = 7 + \frac{3}{8} = \frac{7 \cdot 8}{8} + \frac{3}{8} = \frac{56}{8} + \frac{3}{8} = \frac{59}{8}$$

$$27\frac{5}{11} = 27 + \frac{5}{11} = \frac{27 \cdot 11}{11} + \frac{5}{11} = \frac{302}{11}$$

To convert a positive mixed number to an ordinary fraction, multiply the whole number part by the denominator of the fraction part; then add the result to the numerator of the fraction part to obtain the numerator of the new number.

Check that the following pairs are equivalent.

$$\frac{9}{4} = 2\frac{1}{4} \qquad \frac{8}{3} = 2\frac{2}{3}$$

$$\frac{8}{5} = 1\frac{3}{5} \qquad \frac{20}{3} = 6\frac{2}{3}$$

$$\frac{10}{7} = 1\frac{3}{7} \qquad \frac{35}{2} = 17\frac{1}{2}$$

Any fraction that has 0 as its numerator is equal to 0. For example, $\frac{0}{4} = 0$ and $\frac{0}{7} = 0$. Using the models at the beginning of the chapter, try to explain why this is so. **Zero is never the denominator** of a fraction; the reason that zero never appears as a denominator is that we cannot divide an integer by 0 and get another number; for example, if $6 \div 0 = n$ then $n \cdot 0 = 6$, but $n \cdot 0$ is always 0.

Exercises D-2

I. List 5 fractions that are equivalent to each of the following.

1. 2

2. 70

3. $\frac{3}{2}$

4. $\frac{2}{5}$

5. $\frac{4}{7}$

6. $\frac{3}{8}$

7. 5

8. $\frac{5}{11}$

9. $\frac{7}{9}$

10. $\frac{49}{50}$

11. $\frac{25}{32}$

12. $\frac{32}{50}$

II. Change each mixed number to a fraction.

1. $2\frac{1}{3}$

2. $1\frac{7}{8}$

3. $5\frac{2}{5}$

4. $3\frac{3}{4}$

5. $4\frac{1}{8}$

6. $1\frac{3}{4}$

7. $2\frac{1}{3}$

8. $3\frac{1}{4}$

9. $2\frac{7}{8}$

10. $13\frac{3}{10}$

III. Change each fraction to a mixed number.

1. $\frac{7}{5}$

2. $\frac{16}{5}$

3. $\dfrac{24}{12}$

4. $\dfrac{95}{8}$

5. $\dfrac{121}{25}$

6. $\dfrac{25}{3}$

7. $\dfrac{17}{4}$

8. $\dfrac{29}{5}$

9. $\dfrac{32}{12}$

10. $\dfrac{101}{11}$

D.3 ORDER OF FRACTIONS

The principle of trichotomy that applies to positive whole numbers (see Chapter 3) and to integers (see Chapter 4) applies to fractions as well.

PRINCIPLE OF TRICHOTOMY

If $\dfrac{a}{b}$ and $\dfrac{c}{d}$ are any two fractions, then exactly one of the following is true.

$$\frac{a}{b} < \frac{c}{d}$$

$$\frac{a}{b} = \frac{c}{d}$$

$$\frac{a}{b} > \frac{c}{d}$$

In Chapter 4 we observed that, for integers, we can interpret $a < b$ to mean a lies to the left of b on the number line; similarly, we can interpret $a > b$ to mean a lies to the right of b. The same interpretation holds true for fractions. We can use this interpretation together with our number line model to develop a rule for determining which of the three relationships $(<, =, >)$ holds for any pair of fractions.

Consider the fractions $\dfrac{11}{8}, \dfrac{7}{8}, \dfrac{4}{8}$, and $\dfrac{3}{8}$, each with the denominator 8. If we place these fractions on the number line (Figure D.13), we see that $\dfrac{3}{8} < \dfrac{4}{8} < \dfrac{7}{8} < \dfrac{11}{8}$, and these fractions are in the same order as their numerators $(3 < 4 < 7 < 11)$. A similar relationship occurs whenever we compare any two fractions with the **same** denominator; that is,

Figure D.13

$$\frac{a}{d} < \frac{c}{d} \qquad \text{if } a < c$$

$$\frac{a}{d} = \frac{c}{d} \qquad \text{if } a = c$$

$$\frac{a}{d} > \frac{c}{d} \qquad \text{if } a > c$$

Suppose we wish to compare two fractions with **different** denominators; for example, $\frac{2}{3}$ and $\frac{3}{4}$. We know from our work in the last section, that each of these fractions has many equivalent forms.

$$\frac{2}{3} = \frac{4}{6} = \frac{6}{9} = \frac{8}{12} = \frac{10}{15} = \cdots$$

$$\frac{3}{4} = \frac{6}{8} = \frac{9}{12} = \frac{12}{16} = \frac{15}{20} = \cdots$$

Instead of comparing $\frac{2}{3}$ directly with $\frac{3}{4}$, we can find equivalent forms of both of these fractions with the same denominator and compare them using the rule above. Thus, since $\frac{2}{3} = \frac{8}{12}$ and $\frac{3}{4} = \frac{9}{12}$, we compare $\frac{8}{12}$ and $\frac{9}{12}$. Since $8 < 9$, $\frac{8}{12} < \frac{9}{12}$; consequently, $\frac{2}{3} < \frac{3}{4}$. We apply the same method to compare $\frac{7}{8}$ with $\frac{5}{6}$

$$\frac{7}{8} = \frac{7 \cdot 6}{8 \cdot 6} = \frac{42}{48}$$

$$\frac{5}{6} = \frac{5 \cdot 8}{6 \cdot 8} = \frac{40}{48}$$

$$\frac{42}{48} > \frac{40}{48} \qquad \text{since } 42 > 40$$

Therefore, $\frac{7}{8} > \frac{5}{6}$. Notice that the common denominator we chose in this example was the product $8 \cdot 6$ from the denominators of $\frac{7}{8}$ and $\frac{5}{6}$. This choice of denominator will enable us to compare any two fractions $\frac{a}{b}$ and $\frac{c}{d}$.

$$\text{Compare } \frac{a}{b} \text{ with } \frac{c}{d}.$$

$$\frac{a}{b} = \frac{a \cdot d}{b \cdot d}$$

$$\frac{c}{d} = \frac{b \cdot c}{b \cdot d}$$

Therefore, we can compare the two fractions $\frac{a \cdot d}{b \cdot d}$ and $\frac{b \cdot c}{b \cdot d}$ with a common denominator $b \cdot d$. We already know how to do this by comparing numerators $a \cdot d$ and $b \cdot c$, and we arrive at the following principle.

PRINCIPLE FOR COMPARING FRACTIONS

If $\frac{a}{b}$ and $\frac{c}{d}$ are fractions and a, b, c, and d are positive integers,

$$\frac{a}{b} < \frac{c}{d} \qquad \text{if } a \cdot d < b \cdot c$$

$$\frac{a}{b} = \frac{c}{d} \qquad \text{if } a \cdot d = b \cdot c$$

$$\frac{a}{b} > \frac{c}{d} \qquad \text{if } a \cdot d > b \cdot c$$

You should observe that this principle is an extension of the method we called cross-multiplication, which we used earlier to determine whether two fractions were equal. The following examples illustrate the use of this principle.

EXAMPLE 1   Compare $\frac{8}{13}$ with $\frac{10}{17}$.

To make this comparison consider that $a = 8$, $b = 13$, $c = 10$, and $d = 17$. Then $a \cdot d = 8 \cdot 17 = 136$ and $b \cdot c = 13 \cdot 10 = 130$. Since $136 > 130$, $a \cdot d > b \cdot c$, and it follows that $\frac{8}{13} > \frac{10}{17}$.

EXAMPLE 2 Compare $\frac{15}{24}$ with $\frac{35}{56}$.

Compare $15 \cdot 56$ with $24 \cdot 35$. Computing, $15 \cdot 56 = 840$ and $24 \cdot 35 = 840$. Since $15 \cdot 56 = 24 \cdot 35$, $\frac{15}{24} = \frac{35}{56}$.

Exercises D-C

I. Insert the correct symbol $(<, =, >)$ between the fractions in each pair.

1. $\dfrac{1}{2}$ $\dfrac{3}{8}$

2. $\dfrac{2}{3}$ $\dfrac{3}{5}$

3. $\dfrac{8}{9}$ $\dfrac{17}{18}$

4. $\dfrac{25}{33}$ $\dfrac{17}{25}$

5. $\dfrac{1}{3}$ $\dfrac{3}{8}$

6. $\dfrac{53}{50}$ $\dfrac{26}{25}$

7. $\dfrac{4}{7}$ $\dfrac{14}{17}$

8. $\dfrac{75}{81}$ $\dfrac{25}{27}$

9. $\dfrac{3}{11}$ $\dfrac{8}{33}$

10. $\dfrac{19}{10}$ $\dfrac{39}{20}$

II. Arrange the fractions in each set from least to greatest.

1. $\dfrac{1}{9}$ $\dfrac{1}{11}$ $\dfrac{2}{23}$ $\dfrac{3}{31}$ $\dfrac{2}{10}$

2. $\dfrac{1}{3}$ $\dfrac{1}{4}$ $\dfrac{2}{3}$ $\dfrac{1}{2}$ $\dfrac{3}{4}$

3. $\dfrac{7}{8}$ $\dfrac{17}{18}$ $\dfrac{15}{16}$ $\dfrac{4}{5}$ $\dfrac{21}{22}$

4. $\dfrac{5}{6}$ $\dfrac{4}{3}$ $\dfrac{3}{4}$ $\dfrac{6}{7}$ $\dfrac{2}{3}$

D.4 ARITHMETIC OF FRACTIONS

The rules for adding, subtracting, multiplying, and dividing fractions follow from the meaning of these operations as applied to these numbers. In this section we examine the rules for the operations together with their meanings.

MULTIPLICATION

Multiplication is a simple operation on fractions. To multiply two fractions, we multiply the numerators to find the numerator of the product and multiply the denominators to find the denominator of the product. Here are some examples.

$$\frac{2}{3} \cdot \frac{5}{8} = \frac{2 \cdot 5}{3 \cdot 8} = \frac{10}{24}$$

$$\frac{5}{6} \cdot \frac{4}{7} = \frac{5 \cdot 4}{6 \cdot 7} = \frac{20}{42}$$

$$\frac{1}{3} \cdot \frac{11}{4} = \frac{1 \cdot 11}{3 \cdot 4} = \frac{11}{12}$$

$$\frac{2}{3} \cdot \frac{4}{5} = \frac{2 \cdot 4}{3 \cdot 5} = \frac{8}{15}$$

To see why this is the appropriate way to multiply fractions, we need to refer to the meaning of multiplication and the meaning of fractions.

One meaning for multiplication of whole numbers which we have used frequently in this text is related to the area of a rectangle. For example, $8 \cdot 5$ is the number of square units needed to cover a rectangle that is 8 (linear) units long and 5 units wide. Take a look at Figure D.14.

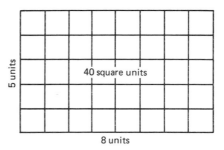

Figure D.14

Now suppose we have a rectangle that is $\frac{3}{4}$ unit long and $\frac{1}{2}$ unit wide.

Its area should be $\left(\dfrac{3}{4}\right) \cdot \left(\dfrac{1}{2}\right)$. We examine such a rectangle in Figure D.15; its area is clearly less than 1 square unit.

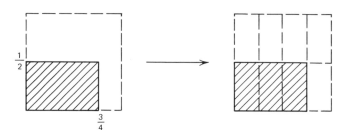

Figure D.15

What part of a square unit is illustrated in Figure D.15? To answer this question, we can divide the square unit into $4 \cdot 2$ identical pieces as shown, and observe that the rectangle consists of $3 \cdot 1$ of these pieces. Thus, the rectangle comprises $\dfrac{3}{8}$ of the unit square and its area is $\dfrac{3}{8}$; that is, $\dfrac{3}{4} \cdot \dfrac{1}{2} = \dfrac{3 \cdot 1}{4 \cdot 2} = \dfrac{3}{8}$. Examine the following examples that show further illustrations of multiplication.

EXAMPLE 1 $\dfrac{4}{5} \cdot \dfrac{3}{8}$.

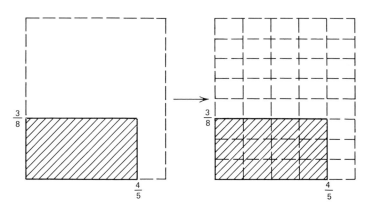

The square unit consists of $5 \cdot 8 = 40$ identical parts; $4 \cdot 3 = 12$ of these comprise the rectangle. Therefore, the area of the rectangle is $\dfrac{4}{5} \cdot \dfrac{3}{8} = \dfrac{4 \cdot 3}{5 \cdot 8} = \dfrac{12}{40}$.

EXAMPLE 2 $\dfrac{3}{4} \cdot \dfrac{3}{2}$.

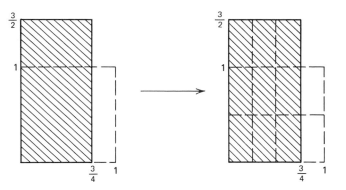

The unit square consists of $4 \cdot 2 = 8$ identical parts; the rectangle consists of $3 \cdot 3 = 9$ parts identical to these; thus, the rectangle has an area equal to $1\dfrac{1}{8}$ or $\dfrac{9}{8}$ square units and $\dfrac{3}{4} \cdot \dfrac{3}{2} = \dfrac{3 \cdot 3}{4 \cdot 2} = \dfrac{9}{8}$. This example leads us to the following definition.

DEFINITION OF MULTIPLICATION OF FRACTIONS

$$\frac{a}{b} \cdot \frac{c}{d} = \frac{a \cdot c}{b \cdot d}$$

DIVISION

When we think of division of natural numbers, for example $12 \div 3$, we generally think of a question such as "How many 3's are there in 12?" Similarly, the quotient $\dfrac{3}{4} \div \dfrac{5}{7}$ can be considered in relation to the question, "How many $\dfrac{5}{7}$'s are there in $\dfrac{3}{4}$?" Before attacking this particular fraction division, we examine some easier examples.

EXAMPLES 1 $\dfrac{3}{4} \div \dfrac{1}{4}$ or how many $\dfrac{1}{4}$'s are there in $\dfrac{3}{4}$?

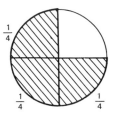

$$\frac{3}{4} \div \frac{1}{4} = 3 \div 1 = \frac{3}{1} = 3$$

EXAMPLE 2 $\dfrac{2}{3} \div \dfrac{1}{6}$ or how many $\dfrac{1}{6}$'s are there in $\dfrac{2}{3}$?

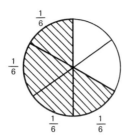

$\dfrac{2}{3}$ is equivalent to $\dfrac{4}{6}$

$$\frac{2}{3} \div \frac{1}{6} = \frac{4}{6} \div \frac{1}{6} = 4 \div 1 = \frac{4}{1} = 4$$

EXAMPLE 3 $\dfrac{3}{8} \div \dfrac{3}{4}$ or how many $\dfrac{3}{4}$'s are there in $\dfrac{3}{8}$?

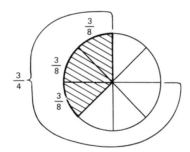

$\dfrac{3}{4}$ is equivalent to $\dfrac{6}{8}$

$$\frac{3}{8} \div \frac{3}{4} = \frac{3}{8} \div \frac{6}{8} = 3 \div 6 = \frac{3}{6} = \frac{1}{2}$$

In each of these examples we have found a common denominator for the fractions and then divided the numerators. If we apply this method to our original example of $\dfrac{3}{4} \div \dfrac{5}{7}$, we obtain the following.

$$\frac{3}{4} \div \frac{5}{7} = \frac{3 \cdot 7}{4 \cdot 7} \div \frac{4 \cdot 5}{4 \cdot 7} = (3 \cdot 7) \div (4 \cdot 5)$$

$$= \frac{3 \cdot 7}{4 \cdot 5} = \frac{21}{20}$$

This computation indicates that the number of $\dfrac{5}{7}$'s in $\dfrac{3}{4}$ is $\dfrac{21}{20}$. Here is another example.

$$\frac{5}{8} \div \frac{2}{3} = \frac{5 \cdot 3}{8 \cdot 3} \div \frac{8 \cdot 2}{8 \cdot 3} = (5 \cdot 3) \div (8 \cdot 2)$$

$$= \frac{5 \cdot 3}{8 \cdot 2} = \frac{15}{16}$$

A similar procedure will work for any pair of fractions.

If we repeat this procedure for arbitrary fractions $\frac{a}{b}$ and $\frac{c}{d}$, we obtain the following.

$$\frac{a}{b} \div \frac{c}{d} = \frac{a \cdot d}{b \cdot d} \div \frac{b \cdot c}{b \cdot d} = (a \cdot d) \div (b \cdot c)$$

$$\frac{a}{b} \div \frac{c}{d} = \frac{a \cdot d}{b \cdot c}$$

Notice that we do not need to find a common denominator for division of fractions. We can shorten this computational procedure by observing that $\frac{a \cdot d}{b \cdot c} = \frac{a}{b} \cdot \frac{d}{c}$. Thus,

$$\frac{a}{b} \div \frac{c}{d} = \frac{a}{b} \cdot \frac{d}{c}$$

We summarize our observations in this definition.

DEFINITION OF DIVISION OF FRACTIONS

$$\frac{a}{b} \div \frac{c}{d} = \frac{a}{b} \cdot \frac{d}{c} = \frac{a \cdot d}{b \cdot c}$$

You may remember a rule for dividing fractions described as **invert and multiply**. Notice that the definition of division accomplishes just that; the divisor (the second fraction) is inverted and multiplied by the first fraction.

ADDITION

How should we add two fractions? Just as our definitions for multiplication and division of fractions had to conform to our ideas about the meanings of the operations for whole numbers and integers, our addition principle must agree with the meaning of addition for these numbers. A very natural way to try to add is to imitate our multiplication rule.

$$\frac{1}{2} + \frac{3}{8} \rightarrow \frac{1 + 3}{2 + 8} = \frac{4}{10}$$

Unfortunately, there are several problems with this "rule". One problem is revealed by examining the sum in comparison with the fractions being added. Take a look at Figure D.16.

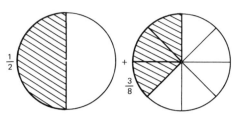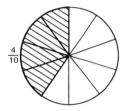

Figure D.16

$\frac{4}{10}$ is less than $\frac{1}{2}$! When we add positive numbers, we should arrive at a sum that is larger than either of them.

A second problem with this "rule" is that we can arrive at different answers for the same problem, depending on which form of a fraction we use. For example, since $\frac{1}{2} = \frac{2}{4}$, let us "add" $\frac{2}{4}$ and $\frac{3}{8}$.

$$\frac{2}{4} + \frac{3}{8} \rightarrow \frac{2+3}{4+8} = \frac{5}{12}$$

This answer is different from the one obtained above. For these reasons this cannot be our rule for addition, and we must seek another one.

As we did for division we begin with fractions having the same denominator. It is not too difficult to see how these should be added using our number line model. Consider these examples.

EXAMPLE 1 $\frac{7}{8} + \frac{3}{8}$.

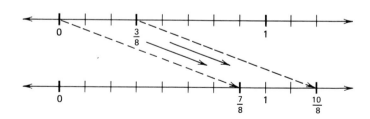

First find $\frac{7}{8}$; the sum corresponds to the point that is $\frac{3}{8}$ unit to the right of $\frac{7}{8}$, or $\frac{7}{8} + \frac{3}{8} = \frac{10}{8}$.

EXAMPLE 2 $\dfrac{1}{5} + \dfrac{3}{5}$.

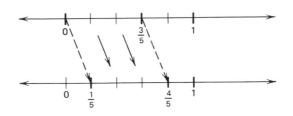

$$\frac{1}{5} + \frac{3}{5} = \frac{4}{5}$$

EXAMPLE 3 $\dfrac{4}{8} + \dfrac{3}{8}$.

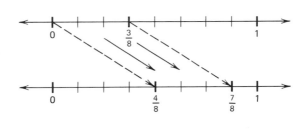

$$\frac{4}{8} + \frac{3}{8} = \frac{7}{8}$$

As these examples suggest, two fractions with the same denominator can be added by adding their numerators; that is, $\dfrac{a}{b} + \dfrac{c}{b} = \dfrac{a+c}{b}$.

In general, we want to be able to add any two fractions. The definition above (and our experience comparing fractions) suggests that to add two fractions we must first convert them to equivalent forms having the same denominator, and then add them. To return to our original example, $\dfrac{1}{2} + \dfrac{3}{8}$, we can find this sum as follows.

$$\frac{1}{2} + \frac{3}{8} = \frac{4}{8} + \frac{3}{8} = \frac{4+3}{8} = \frac{7}{8}$$

There are many denominators that can be used for any two fractions. The addition above could also be performed in the following ways.

$$\frac{1}{2} + \frac{3}{8} = \frac{8}{16} + \frac{6}{16} = \frac{14}{16} = \frac{7}{8}$$

or

$$\frac{1}{2} + \frac{3}{8} = \frac{40}{80} + \frac{30}{80} = \frac{70}{80} = \frac{7}{8}$$

or many other ways. It doesn't matter which number we use as a common denominator as long as both of the fractions to be added can be converted to that denominator. Thus, the common denominator must be a multiple of both denominators. One choice that can *always* work is the product of the two original denominators. We use this choice in the following additional examples.

$$\frac{3}{10} + \frac{5}{8} = \frac{3 \cdot 8}{10 \cdot 8} + \frac{10 \cdot 5}{10 \cdot 8}$$

$$= \frac{24}{80} + \frac{50}{80} = \frac{74}{80} = \frac{37}{40}$$

$$\frac{9}{5} + \frac{7}{6} = \frac{9 \cdot 6}{5 \cdot 6} + \frac{5 \cdot 7}{5 \cdot 6}$$

$$= \frac{54}{30} + \frac{35}{30} = \frac{89}{30} = 2\frac{29}{30}$$

These computations suggest the following general rule.

DEFINITION OF ADDITION OF FRACTIONS

$$\frac{a}{b} + \frac{c}{d} = \frac{a \cdot d + b \cdot c}{b \cdot d}$$

Many readers may be accustomed to a rule for adding fractions that requires finding a *least common denominator,* and thus find the rule above distasteful. The point is that *any* common denominator will work, and the product of the denominators is *always* a common denominator. In other words, the definition above will always work. However, it is not always the most efficient method for adding fractions. Consider this example.

$$\frac{3}{100} + \frac{19}{1000}$$

Using the definition, we would compute

$$\frac{3}{100} + \frac{19}{1000} = \frac{3 \cdot 1000}{100 \cdot 1000} + \frac{19 \cdot 100}{100 \cdot 1000}$$

$$= \frac{3000 + 1900}{100,000} = \frac{4,900}{100,000} = \frac{49}{1000}$$

Using a little common sense (and the least common denominator), we would perform the addition more simply.

$$\frac{3}{100} + \frac{19}{1000} = \frac{30}{1000} + \frac{19}{1000} = \frac{49}{1000}$$

On the other hand, there are times when finding the least common denominator is rather trying, and using the definition is easier. The following computation is not hard on a calculator.

$$\frac{10}{247} + \frac{5}{299} = \frac{10 \cdot 299}{247 \cdot 299} + \frac{247 \cdot 5}{247 \cdot 299} = \frac{2990 + 1235}{73{,}853} = \frac{4225}{73{,}853}$$

Now try it using the "least common denominator" method. You are probably in for a lot of work determining that $247 = 13 \cdot 19$ and $299 = 13 \cdot 23$, and the computation is not much easier.

$$\frac{10}{247} + \frac{5}{299} = \frac{10 \cdot 23}{247 \cdot 23} + \frac{19 \cdot 5}{19 \cdot 299} = \frac{230}{5681} + \frac{95}{5681} = \frac{325}{5681}$$

The message is that when you are adding fractions use an efficient method. If there is an obvious common denominator, use it! If there is not an obvious choice, resort to the definition; it always works.

SUBTRACTION

The rule for subtraction is almost identical to the rule for addition, as is the reasoning behind it. We simply state it in the following way.

DEFINITION OF SUBTRACTION OF FRACTIONS

$$\frac{a}{b} - \frac{c}{d} = \frac{a \cdot d\ b \cdot c}{b \cdot d}$$

As with addition, it is possible to subtract fractions using a denominator other than the product of the denominators.

Exercises D-4

I. Find the following products.

1. $\dfrac{2}{3} \cdot \dfrac{5}{8}$

2. $\dfrac{9}{7} \cdot \dfrac{3}{5}$

3. $\dfrac{8}{5} \cdot \dfrac{5}{8}$

4. $\dfrac{7}{12} \cdot \dfrac{12}{7}$

5. $\dfrac{15}{22} \cdot \dfrac{11}{25}$

6. $\dfrac{9}{100} \cdot \dfrac{50}{3}$

7. $\left(\dfrac{7}{8} \cdot \dfrac{3}{5}\right) \cdot \dfrac{2}{7}$

8. $\dfrac{7}{8} \cdot \left(\dfrac{3}{5} \cdot \dfrac{2}{7}\right)$

9. $\dfrac{5}{6} \cdot \dfrac{8}{10} \cdot \dfrac{7}{12}$

10. $\dfrac{9}{11} \cdot \dfrac{7}{9} \cdot \dfrac{5}{7}$

II. Find the following products. (You will need to change the mixed numbers into ordinary fractions.)

1. $1\dfrac{1}{2} \cdot \dfrac{3}{5}$

2. $2\dfrac{1}{4} \cdot 6$

3. $\dfrac{3}{8} \cdot 1\dfrac{1}{3}$

4. $5 \cdot 2\dfrac{2}{3}$

5. $1\dfrac{1}{2} \cdot 4$

6. $2\dfrac{1}{3} \cdot 3\dfrac{1}{4}$

7. $5\dfrac{1}{2} \cdot 7\dfrac{1}{3}$

8. $9\dfrac{2}{5} \cdot 5\dfrac{1}{10}$

9. $\left(1\dfrac{5}{6}\right)^2$

10. $\left(2\dfrac{1}{2}\right)^3$

III. Find the following quotients.

1. $\dfrac{1}{2} \div \dfrac{3}{4}$

2. $\dfrac{5}{8} \div \dfrac{3}{2}$

3. $\dfrac{7}{8} \div \dfrac{7}{8}$

4. $\dfrac{7}{8} \div \dfrac{8}{7}$

5. $\dfrac{7}{4} \div \dfrac{1}{4}$

6. $\dfrac{15}{4} \div \dfrac{3}{16}$

7. $\dfrac{1}{12} \div \dfrac{1}{2}$

8. $\dfrac{1}{12} \div 2$

9. $3 \div \dfrac{2}{3}$

10. $3 \div \dfrac{3}{2}$

11. $\dfrac{5}{24} \div \dfrac{15}{6}$

12. $\dfrac{9}{11} \div \dfrac{7}{11}$

13. $\dfrac{13}{15} \div \dfrac{6}{25}$

14. $\dfrac{4}{5} \div \dfrac{8}{10}$

15. $\dfrac{32}{99} \div \dfrac{8}{44}$

16. $\dfrac{81}{100} \div \dfrac{9}{10}$

17. $\left(\dfrac{1}{2} \div \dfrac{1}{3}\right) \div \dfrac{2}{5}$

18. $\dfrac{1}{2} \div \left(\dfrac{1}{3} \div \dfrac{2}{5}\right)$

19. $\left(\dfrac{3}{8} \div \dfrac{2}{3}\right) \div \dfrac{4}{5}$ **20.** $\dfrac{3}{8} \div \left(\dfrac{2}{3} \div \dfrac{4}{5}\right)$

IV. Compute the following.

1. $\dfrac{1}{2} + \dfrac{2}{3}$ **11.** $\dfrac{2}{3} - \dfrac{1}{2}$

2. $\dfrac{3}{5} + \dfrac{4}{7}$ **12.** $\dfrac{3}{5} - \dfrac{4}{7}$

3. $\dfrac{1}{3} + \dfrac{4}{5}$ **13.** $\dfrac{8}{9} - \dfrac{7}{8}$

4. $\dfrac{7}{24} + \dfrac{9}{16}$ **14.** $\dfrac{5}{12} - \dfrac{1}{8}$

5. $\dfrac{8}{35} + \dfrac{21}{33}$ **15.** $\dfrac{3}{8} - \dfrac{9}{32}$

6. $\dfrac{100}{169} + \dfrac{100}{121}$ **16.** $\dfrac{75}{99} - \dfrac{25}{98}$

7. $1\dfrac{2}{3} + 2\dfrac{1}{2}$ **17.** $2\dfrac{3}{4} - 1\dfrac{1}{2}$

8. $2\dfrac{1}{4} + 3\dfrac{1}{5}$ **18.** $5\dfrac{1}{3} - 4\dfrac{2}{3}$

9. $7\dfrac{1}{8} + 3\dfrac{1}{6}$ **19.** $7\dfrac{1}{4} - 3\dfrac{1}{3}$

10. $9\dfrac{3}{8} + 5\dfrac{5}{12}$ **20.** $6\dfrac{3}{8} - 5\dfrac{7}{12}$

V. Compute the following.

1. $\dfrac{1}{2} \cdot \left(\dfrac{3}{4} + \dfrac{1}{3}\right)$ **4.** $1\dfrac{1}{2} \cdot \left(2\dfrac{1}{3} + 3\dfrac{1}{8}\right)$

2. $\dfrac{2}{3} \cdot \left(\dfrac{5}{7} + \dfrac{1}{2}\right)$ **5.** $2\dfrac{3}{4} \cdot \left(2\dfrac{1}{2} - 1\dfrac{1}{8}\right)$

3. $\dfrac{3}{8} \cdot \left(\dfrac{1}{3} + \dfrac{1}{5}\right)$

VI. Compute the following.

1. $\dfrac{1}{2} \cdot \dfrac{3}{4} + \dfrac{1}{2} \cdot \dfrac{1}{3}$ **4.** $1\dfrac{1}{2} \cdot 2\dfrac{1}{3} + 1\dfrac{1}{2} \cdot 3\dfrac{1}{8}$

2. $\dfrac{2}{3} \cdot \dfrac{5}{7} + \dfrac{2}{3} \cdot \dfrac{1}{2}$ **5.** $2\dfrac{3}{4} \cdot 2\dfrac{1}{2} - 2\dfrac{3}{4} \cdot 1\dfrac{1}{8}$

3. $\dfrac{3}{8} \cdot \dfrac{1}{3} + \dfrac{3}{8} \cdot \dfrac{1}{5}$

VII. Compare your answers in Parts V and VI. What principle is illustrated?

D.5   NEGATIVE FRACTIONS

Consider now the expression $-\frac{3}{4}$. Following our description of negative integers, $-\frac{3}{4}$ should be (and is) the solution to

$$\frac{3}{4} + x = 0$$

That is, $-\frac{3}{4}$ is the number that, when added to $\frac{3}{4}$, gives 0. It follows that $-\frac{3}{4} = \frac{-3}{4}$; to see this observe that

$$\frac{3}{4} + \frac{-3}{4} = \frac{3 + (-3)}{4} \quad \text{(Definition of addition of fractions)}$$

$$= \frac{0}{4} \quad \text{(Meaning of } -3)$$

$$= 0$$

Therefore, $\frac{-3}{4}$ is "the opposite of $\frac{3}{4}$" and $\frac{-3}{4} = -\frac{3}{4}$. Consider now the fraction $\frac{3}{-4}$; like $\frac{-3}{4}$ this is a quotient of two integers, one positive and one negative. To see that $\frac{3}{-4}$ and $\frac{-3}{4}$ are equal, we use the cross-multiplication procedure (see page 000). By cross-multiplying, we compare $3 \cdot 4$ and $(-3) \cdot (-4)$. Since both of these are 12, the fractions are equal.

The principles for comparing negative fractions are analogous to those for comparing negative integers. Recall that if $a < b$, then $-b < -a$. We have a similar rule for fractions; examine the following examples.

$$\frac{1}{4} < \frac{3}{4} \quad \text{and} \quad -\frac{1}{4} > -\frac{3}{4}$$

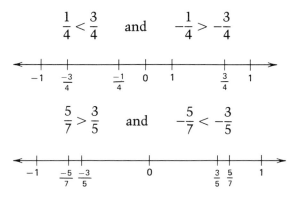

$$\frac{5}{7} > \frac{3}{5} \quad \text{and} \quad -\frac{5}{7} < -\frac{3}{5}$$

Just as we can compare negative fractions by combining our principles for comparing integers with our principles for comparing fractions, we can also perform operations on negative fractions by combining the definitions of operations on fractions with our understanding of operations on negative integers. Consider the following examples.

EXAMPLE 1 $\dfrac{-3}{4} + \dfrac{-2}{3}$.

$$\dfrac{-3}{4} + \dfrac{-2}{3} = \dfrac{-3 \cdot 3 + -2 \cdot 4}{4 \cdot 3} \qquad \text{(Definition of addition of fractions)}$$

$$= \dfrac{-9 + (-8)}{12} \qquad \text{(Multiplication of integers)}$$

$$= \dfrac{-17}{12} \qquad \text{(Addition of integers)}$$

EXAMPLE 2 $-\dfrac{1}{3} - \left(-\dfrac{5}{8}\right)$.

$$-\dfrac{1}{3} - \left(-\dfrac{5}{8}\right) = \dfrac{-1}{3} - \dfrac{-5}{8} \qquad \text{(Meaning of negative fractions)}$$

$$= \dfrac{-1 \cdot 8 - (-5) \cdot 3}{3 \cdot 8} \qquad \text{(Definition of subtraction of fractions)}$$

$$= \dfrac{-8 - (-15)}{24} \qquad \text{(Multiplication of integers)}$$

$$= \dfrac{7}{24} \qquad \text{(Subtraction of integers)}$$

Now practice some computations yourself!

Exercises D-5

Compute the following.

1. $-\dfrac{5}{8} + \dfrac{3}{4}$

2. $\dfrac{-7}{8} + \left(\dfrac{-2}{3}\right)$

3. $\dfrac{3}{4} - \left(-\dfrac{1}{3}\right)$

4. $\dfrac{-3}{4} \cdot \dfrac{5}{-6}$

5. $1\dfrac{4}{5} - \left(-2\dfrac{1}{5}\right)$

6. $\left(-1\dfrac{1}{2}\right) \cdot \left(-2\dfrac{1}{3}\right)$

7. $\left(-\dfrac{5}{8}\right) \div \left(-\dfrac{3}{4}\right)$

8. $\left(-1\dfrac{1}{3}\right) \div \left(-1\dfrac{7}{9}\right)$

9. $\left(\dfrac{5}{8} - \dfrac{7}{12}\right) - \dfrac{3}{4}$

10. $\dfrac{5}{8} - \left(\dfrac{7}{12} - \dfrac{3}{4}\right)$

ANSWERS TO ODD-NUMBERED AND REVIEW PROBLEMS

CHAPTER 1

EXERCISES 1-A, PAGES 5–6

1. 6 hours, 60 miles per hour, 75 miles per hour, $\dfrac{300}{r}$ hours

3. (a)
$$C = \begin{cases} 1.95 & \text{if } t \leq 3 \\ 1.95 + 0.32(t - 3) & \text{if } t > 3 \end{cases}$$

(b) $D = 15n + 25n$

(c) $A = \ell(24 - \ell)$

EXERCISES 1-B, PAGES 9–11

I. 1. $n + 34$　　　　**3.** $5t + 7$　　　　**5.** $k + 9$　　　　**7.** $10 + 7t$

　　9. $(x + 2)^2$　　　**11.** $25t$　　　　**13.** $25t + 10s$　　　**15.** $4x$

II. 1. 10, 19, 4　　　**3.** 15, 55, 105　　**5.** 6, 30, 0　　　**7.** 23, 23, 23

　　9. 36, 16, 100　　**11.** 5, 13, 4　　　**13.** 3, 15, 63

III. 1. 16　　　　　**3.** 10　　　　　**5.** 3　　　　　**7.** 2

　　9. 21

IV. 1. (a) 552   **(b)** 408   **(c)** 462
  3. $37N + 26(20 - N)$

EXERCISES 1-C, PAGES 13–15

I. 1. Expression   **3.** Equation   **5.** Expression
II. 1. none are solutions   **3.** 2 and 3 are solutions
  5. $x = 3, y = 4$ and $x = 2, y = 0$ are solutions
III. 1. $y = 2x + 1$ 2-3. $y = 2^x$ 3.5. $y = 5x + 2$
IV. 1. 5   **3.** 4 rolls of quarters, 8 rolls of nickels
  5. \$12.80

REVIEW PROBLEMS, PAGES 15–16

1. (a) 51, 26, 1   **(b)** 140, 32, 0   **(c)** 136, 16, 6
2. (a) 31   **(b)** 77
3. (a) $x = 4$ **(b)** $x = 4, x = 5$ **(c)** $y = 1$
4. (a) The second number is 4 times the first number: $y = 4x$.
  (b) The second number is 3 more than twice the first: $y = 2x + 3$.
5. \$176
6. 150 bags

CHAPTER 2

EXERCISES 2-A, PAGE 20

I. 1. 4   **3.** 9   **5.** 2   **7.** 4   **9.** 3
II. 1. 2   **3.** 2   **5.** 3
III. 1. (a) no   **(b)** yes, $a = 0, b = 1$
  3. (a) $30 \cdot 2^6, 30 \cdot 2^{10}, 30 \cdot 2^{20}, 30 \cdot 2^{2n}$
  (b) 2 hours

EXERCISES 2-B, PAGE 25

I. 1. $2^4 \cdot 3^2 \cdot 5$ **3.** 10^8 **5.** 3^9 **7.** $3^5 \cdot 7^4$
  9. 2^{20}
II. 1. $x^2 \cdot 4x$ **3.** $x^2 \cdot 141x^3y$ **5.** $x^2 \cdot x^3t^5$
III. 1. $8x^3$ **3.** $6x^7$ **5.** $5x^6$ **7.** $8x^{10}$ **9.** $10y^7$
  11. $81z^{11}$ **13.** $16x^8y^4$ **15.** $36x^{12}$ **17.** $6400t^4x^{18}y^4$
  19. $8x^8$ **21.** $4a^8b^2$ **23.** $8x^9$

EXERCISES 2-C, PAGES 27–28

I. **1.** $7x + 28$

    7. $21x + 21$

    13. $2x^3 + 3x^2 + x$

    19. $2xy + xz + yz$

    3. $k^2 + 9k + 14$

    9. $2x^2 + 12x + 13$

    15. $x^4 + 2x^3 + x^2$

    21. $2ab + 2ac + 2bc$

    5. $2t^2 + 11t + 15$

    11. $4t^2 + 12t + 9$

    17. $4x^4$

    23. $9y^6 + 18y^5 + 9y^4 + 6y^3$

II. **1.** \$6204   **3.** \$47.73   **5.** 400 square feet, 450 square feet, 450 square feet

III. answers in book

REVIEW PROBLEMS, PAGE 29

1. (a) 4   (b) 5

2. (a) $3^3x^4 = 27x^4$

   (c) $4^5x^5 = 1024x^5$

   (b) $2^4t^{12} = 16t^{12}$

   (d) $3^3x^7 = 27x^7$

3. (a) $x^2 + 6x^3$

   (c) $4x^2 + 20x + 25$

   (b) $10y^2 - y$

   (d) $17^2x^4 = 289x^4$

4. \$9000, \$18,000, \$72,000

5. $5[.19(200) + 65] = \$515$

CHAPTER 3

EXERCISES 3-A, PAGES 37–38

I. **1.** 3   **3.** 4   **5.** 3   **7.** 6   **9.** 15

   11. 6   **13.** 7   **15.** 3   **17.** 4   **19.** 19

   21. 11   **23.** 7   **25.** no solution

II. **1.** $\ell = \dfrac{p}{2} - w$ **3.** $b = c - md$ **5.** $w = z$

III. **1.** $1.14 = .11p + .15$, p = number of pencils

    3. $40 = 2((2 + w) + w)$, w = width

    5. $165 = 55t$, t = time

    7. $60t = 80(t - 2)$

EXERCISES 3-B, PAGE 41

1. $x < 5$ **3.** $x \le 10$ **5.** $x < 7$ **7.** $x > 2$

9. $x \ge 5$ **11.** any number   **13.** $2 > x$ **15.** $x < 1$

17. any number   **19.** $0 \le x$ **21.** $8 > x$ **23.** $x \ge 31$

EXERCISES 3-C, PAGES 43–44

I. (Exercises 3-A)

   1.

   3.

5.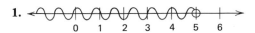

(Exercises 3-B)

1.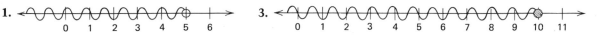

3.

5.

II. 1. 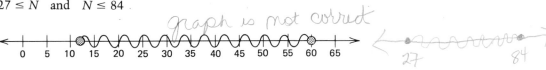   **3.**

III. 1. $27 \leq N$ and $N \leq 84$

graph is not correct

3·C

    3. $6 \leq x$ and $x \leq 9.50$

    5. Mrs. Smith wrote 28 checks and Mr. Smith wrote 7

    7. Between $9\frac{1}{2}$ and $10\frac{1}{2}$ hours $\left(\text{including } 10\frac{1}{2} \text{ but not } 9\frac{1}{2}\right)$

IV. Answers in chapter

REVIEW PROBLEMS, PAGE 45

1. (a) $x = 9$ **(b)** $y = 7$ **(c)** $t = 2$
   (d) $x = 12$ **(e)** $x = 2$ **(f)** $x = 1$

2. (a) $b = c - 2a$ **(b)** $t = \dfrac{d}{r}$

3. (a) $x > 2$ **(b)** $1 < x$

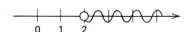   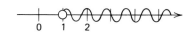

4. 7   **5.** 2 hours, 1 hour

CHAPTER 4

EXERCISES 4-A, PAGE 52

  1. -3 **3.** -4 **5.** -5 **7.** -400 **9.** -35
11. -3 **13.** -15 **15.** -15

EXERCISES 4-B, PAGES 53–54

I. **1.** 6 **3.** −5 **5.** −25 **7.** −3 **9.** −6
 11. 6 **13.** 3 **15.** 19 **17.** −4 **19.** −2

II. February: 9 inches
  March: 15 inches and −4 inches
  April: 7 inches and 22 inches
  May: 32 inches and −1 inches

EXERCISES 4-C, PAGES 55–56

I. **1.** 13 **3.** 15 **5.** −23 **7.** −24 **9.** −12
 11. −18 **13.** 4 **15.** 17

II. **1.** −5 **3.** 2 **5.** 0 **7.** 18 **9.** −9
 11. −2 **13.** −4 **15.** 4 **17.** 16 **19.** 9
 21. 81 **23.** −125

EXERCISES 4-D, PAGES 57–58

I. **1.** 8, 2, 0, 2, 8 **3.** 22, 2, 0, −2, −22
 5. −24, −2, 0, −90 **7.** −6, −1, 4, 9, 14

II. **1.** $1 + x^4$ **3.** $-32x^5$ **5.** 0
 7. $-432x^{10}$ **9.** $-81x^4$

III. **1.** $6x^3 - 15x^2$ **3.** $-125x^3 + 125x^5$ **5.** $15x^3 + 35x$
 7. $-2x^3 + 12x - 6x^2$ **9.** $7 - 3x$

EXERCISES 4-E, PAGE 59

I. **1.** < **3.** > **5.** < **7.** < **9.** =

II. **1.**

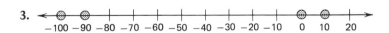

3.

III. **1.**

3.

5.

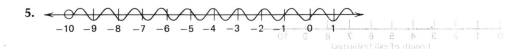

EXERCISES 4-F, PAGES 62–63

I. 1. $x > -4$ **3.** $1 > x$ **5.** $1 < x$ **7.** $-\dfrac{7}{2} < x$

9. $x < -\dfrac{5}{2}$

II. 1. 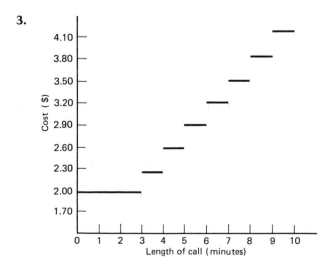   **3.**

5. No solution

III. 1. 30.82, 14.85, -16.54, -29.12, -69.53, -62.52, -39.88, 16.86, 22.25, 28.63, 37.60, 38.09
   They owe $28.49

REVIEW PROBLEMS, PAGE 64

1. (a) 12   **(b)** 20   **(c)** -27 **(d)** -18

2. 33, 5   **3. (a)** $-3x^7$ **(b)** $8x - 2x^2$

4.

5. (a) $-3 < x$ **(b)** $x < -3$

6. 784.54

7. $10 loss, $10, $70

CHAPTER 5

EXERCISES 5-A, PAGES 70–72

1.   (a) Approximately $1200   **(b)** 6 years

3.

5.

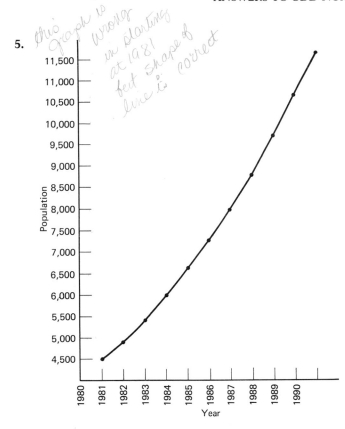

this graph is wrong in starting at 1981 but shape of line is correct

7.

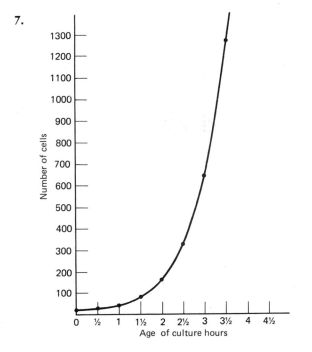

EXERCISES 5-B, PAGES 76–79

1. **(a)** $(2, 1)$ **(b)** $(0, 5)$ **(c)** $(-2, 3)$ **(d)** $(-8, 1)$
     (e) $(-4, -2)$ **(f)** $(-1, -6)$ **(g)** $(4, -3)$ **(h)** $(1, 0)$
     (i) $(7, 2)$

3.

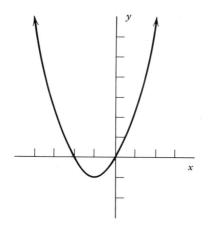

5.

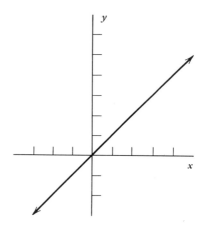

7. $y = 2x + 4$

9.

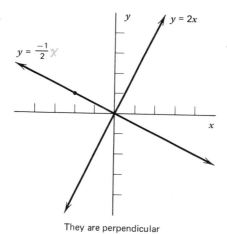

They are perpendicular

11.

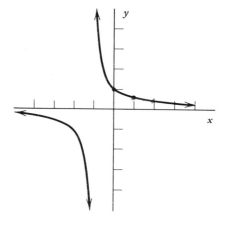

13. **(a)** $0, 3, -1, 0, -\dfrac{1}{2}, 1, 8, 8$

     (b) $1, 0$ and $2, -\dfrac{1}{2}, -1$ and 3

15. (a)

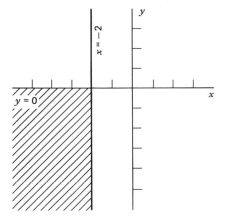

(b)

should be x > 0

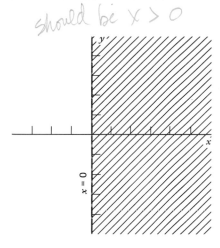

(c)

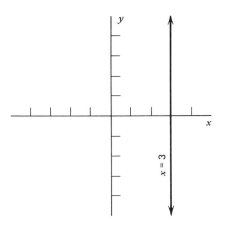

(d)

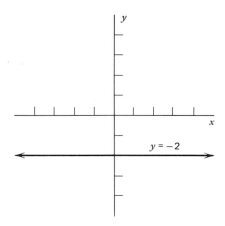

17. (a)

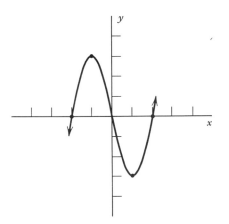

(b)

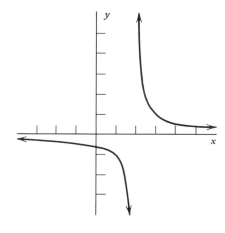

19. (a)

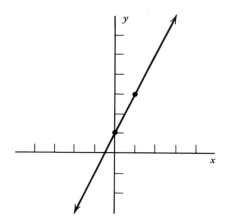

(b)

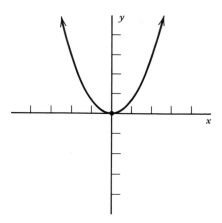

(c)

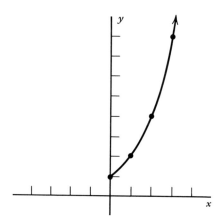

(d)

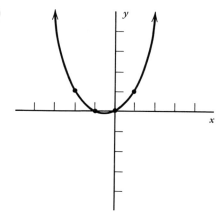

(e)

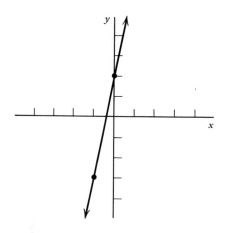

REVIEW PROBLEMS, PAGE 79

1.

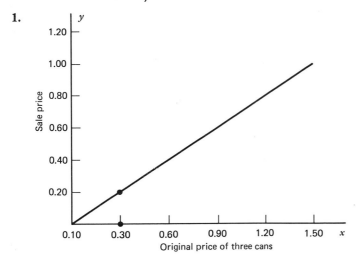

2.

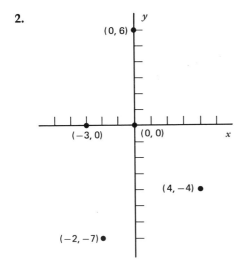

3. $y = 3x^2$

Sample Solutions:

$x = 0,\ y = 0$

$x = 1,\ y = 3$

$x = -1,\ y = 3$

$x = 2,\ y = 12$

$x = -2,\ y = 12$

$x = \dfrac{1}{3},\ y = \dfrac{1}{3}$

$x = -\dfrac{1}{3},\ y = \dfrac{1}{3}$

$x = \dfrac{2}{3},\ y = \dfrac{4}{3}$

$x = -\dfrac{2}{3},\ y = \dfrac{4}{3}$

$x = 3,\ y = 27$

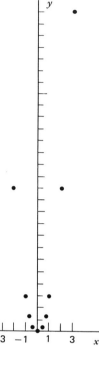

4. $y = 4x + 3$

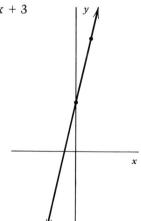

5. $y = \dfrac{1}{2x}$

10 solutions:

$x = 1,\ y = \dfrac{1}{2}$ $x = -1,\ y = -\dfrac{1}{2}$

$x = 2,\ y = \dfrac{1}{4}$ $x = -2,\ y = -\dfrac{1}{4}$

$x = 8,\ y = \dfrac{1}{16}$ $x = -10,\ y = -\dfrac{1}{20}$

$x = \dfrac{1}{2},\ y = 1$ $x = -\dfrac{1}{2},\ y = -1$

$x = \dfrac{1}{10},\ y = 5$ $x = -\dfrac{1}{8},\ y = -4$

5. continued

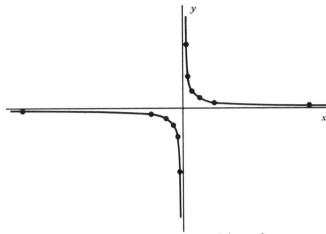

There is no point on the graph for $x = 0$

CHAPTER 6

EXERCISES 6-A, PAGE 85

I. **1.** slope: 3   intercept: 1
   3. slope: -3 intercept: 4
   5. slope: -4 intercept: 0
   7. slope: undefined   intercept: none
   9. slope: $-\dfrac{1}{4}$ intercept: 2

II. **1.** y-intercept: 2 x-intercept: 3   **3.** y-intercept: 3 x-intercept: -6

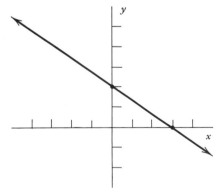

III. **1.** $y = 2x + 1$ **3.** $y = \dfrac{1}{2}x$ **5.** $y = -9x - 3$

   7. $y = 4$ **9.** $x = \dfrac{3}{2}$ **11.** $y = 0$

EXERCISES 6-B, PAGES 89–91

I. 1.

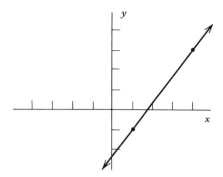

3.

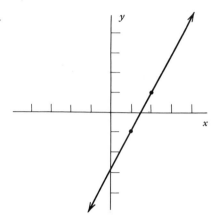

5.

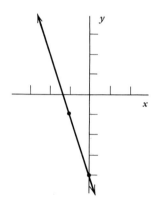

7.

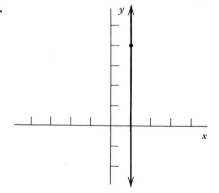

II. 1.

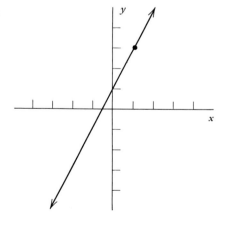

3.

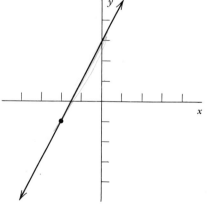

5.

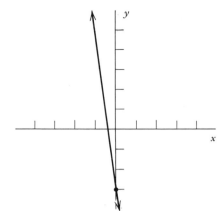

7.

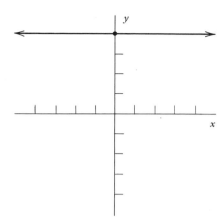

9.

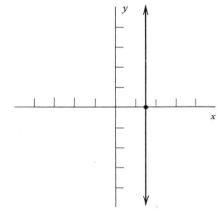

11.

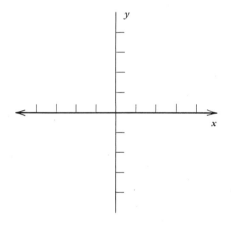

III. **1.** 1   **3.** $-\dfrac{4}{7}$ **5.** $-\dfrac{3}{2}$ **7.** 0   **9.** -2

11. $-\dfrac{1}{2}$

IV. **1.** There is division by zero

V. **1.** $y = \dfrac{1}{2}x + 5$ **3.** $y = 3$ **5.** $y = -\dfrac{2}{3}x - \dfrac{13}{3}$

     7. $y = 10x - 40$ **9.** $y = \dfrac{4}{3}x + \dfrac{20}{3}$

VI. 1. $y = \dfrac{1}{3}x + \dfrac{2}{3}$

3. $y = -\dfrac{3}{2}x - \dfrac{17}{2}$

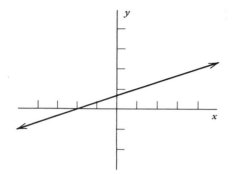

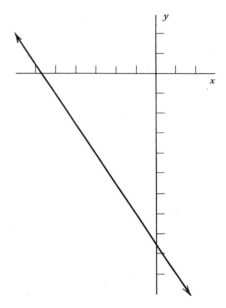

5. $y = -x + 2$

7. $y = 1$

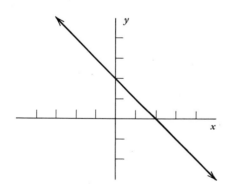

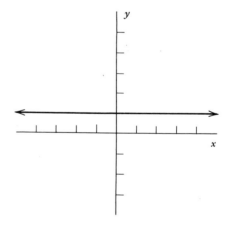

9. $y = -\dfrac{2}{5}x + \dfrac{8}{5}$

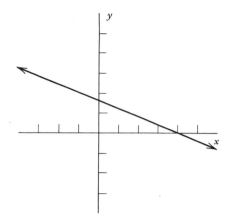

VII. 1. $F = \dfrac{9}{5}C + 32$ slope $= \dfrac{9}{5}$ y-intercept $= 32$

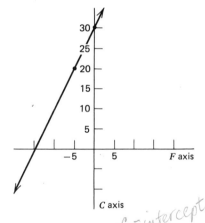

3. $c = 0.03f + 120$; slope $= 0.03$; y-intercept $= 120$

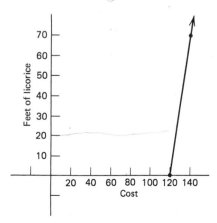

EXERCISES 6-C, PAGES 96–97

I. **1.** $x = 5, y = 2$ **3.** No solution   **5.** $x = 5, y = 2$
   7. $x = 7, y = 2$ **9.** $x = 0, y = 0$ **11.** $x = 4, y = -15$
   13. $x = -1, y = \dfrac{1}{2}$ **15.** $x = 2, y = 3$

II. **1.** No solution

III. **1.** Nuts cost $2.00 per pound and bolts $3.00 per pound
   3. 22 fifteen-cent stamps and 3 eight-cent stamps
   5. $w = 50$ miles per hour and $s = 350$ miles per hour
   7. 22.3 ounces of gin and 9.7 ounces of vermouth

EXERCISES 6-D, PAGES 100–101

I. **1.** $x = 4, y = 2, z = -3$ **3.** $x = 5, y = 0, z = -2$
   5. $x = -\dfrac{9}{2}, y = \dfrac{5}{4}, z = 3$

II. **1.**   30 pounds of Himalayan tea
        50 pounds of Indian tea
       120 pounds of domestic tea

REVIEW PROBLEMS, PAGES 101–102

1. (a) slope: -2 **(b)** slope: $-\dfrac{1}{3}$

   y-intercept: 1 y-intercept: 3

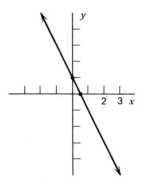
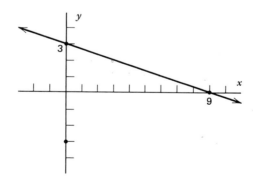

(c) slope: 0

y-intercept: 3

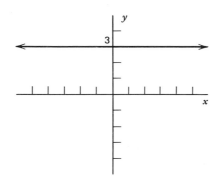

2. x-intercept: 2

y-intercept: −3

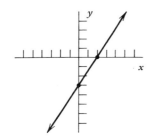

3. $\dfrac{5}{2}$

4. (a) $y = \dfrac{1}{2}x - 1$ **(b)** $y = -\dfrac{4}{3}x + \dfrac{14}{3}$

   (c) $y = x - 5$ **(d)** $x - 2y = -2$

   (e) $x = 3$ **(f)** $y = -5$

5. (a) $x = 4, y = 2$ **(b)** $x = -3, y = 2$

   (c) no solutions

6. $x = 2, y = 0, z = 2$

7. 280 lbs sausage, 670 lbs cheese

8. 100 lbs hamburger, 50 lbs soybean meal

9. $T = .0055 \,(.35\,V)$, \$144.38

CHAPTER 7

EXERCISES 7-A, PAGES 108–109

I.

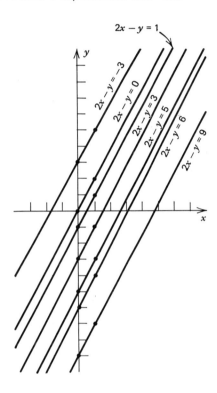

7. The lines are all parallel to $2x - y = 5$.

Pairs satisfying the last two equations satisfy the condition $2x - y > 5$.

Pairs satisfying the first four equations satisfy the condition $2x - y < 5$.

II. 1.

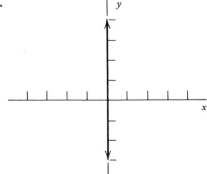

3.

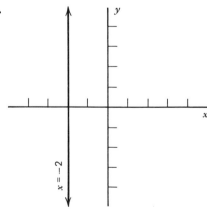

5.

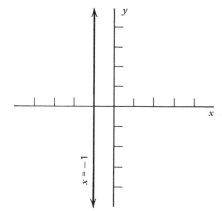

III. 1.

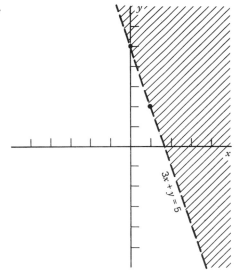

3.

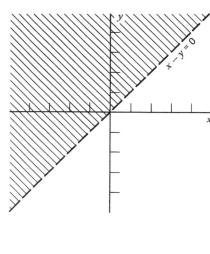

5.

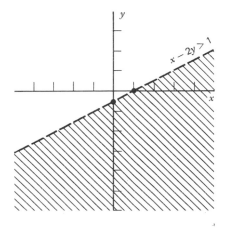

7.

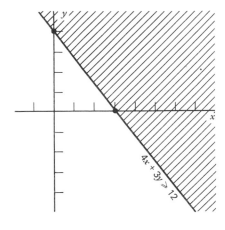

9.

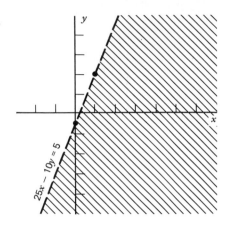

11.

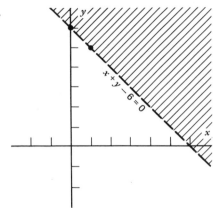

13.

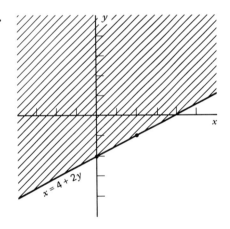

15.

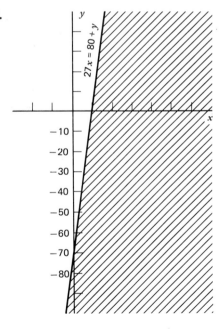

17.

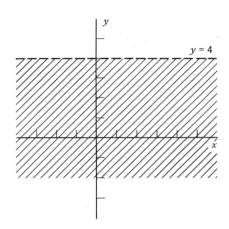

19.

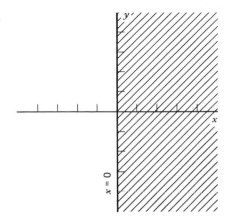

EXERCISES 7-B, PAGES 113–114

I. 1.

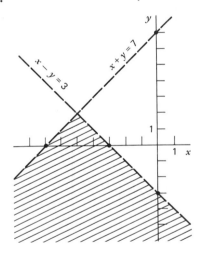

3.

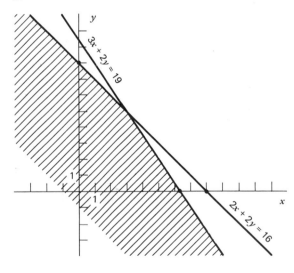

5.

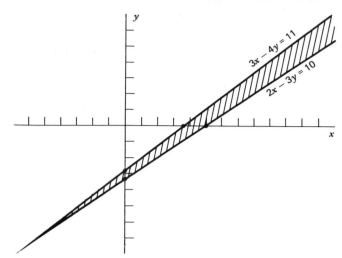

7.

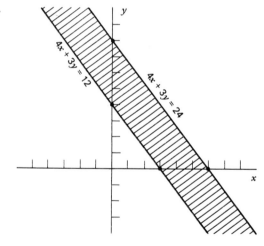

9. no solution

II. 1.

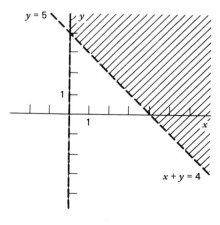

3.

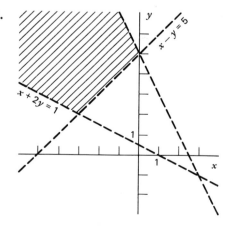

5.

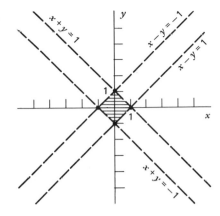

EXERCISES 7-C, PAGE 116

1.

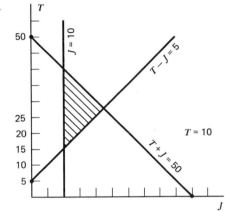

EXERCISES 7-D, PAGES 121–123

I. 1. E   **3.** A   **5.** E   **7.** D

II. 1.

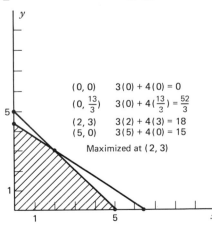

$(0, 0)$ $3(0) + 4(0) = 0$

$(0, \frac{13}{3})$ $3(0) + 4(\frac{13}{3}) = \frac{52}{3}$

$(2, 3)$ $3(2) + 4(3) = 18$

$(5, 0)$ $3(5) + 4(0) = 15$

Maximized at $(2, 3)$

3.

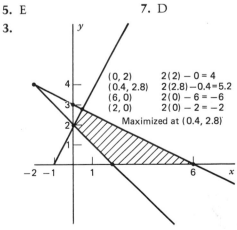

$(0, 2)$ $2(2) - 0 = 4$

$(0.4, 2.8)$ $2(2.8) - 0.4 = 5.2$

$(6, 0)$ $2(0) - 6 = -6$

$(2, 0)$ $2(0) - 2 = -2$

Maximized at $(0.4, 2.8)$

5.

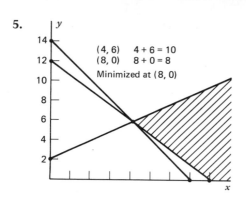

(4, 6)   4 + 6 = 10
(8, 0)   8 + 0 = 8
Minimized at (8, 0)

III. 1.

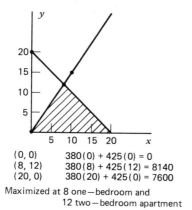

(0, 0)   380(0) + 425(0) = 0
(8, 12)   380(8) + 425(12) = 8140
(20, 0)   380(20) + 425(0) = 7600

Maximized at 8 one—bedroom and
12 two—bedroom apartment

3.

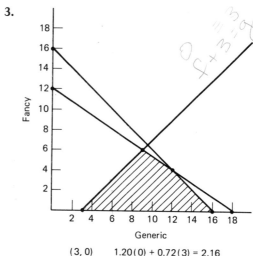

(3, 0)   1.20(0) + 0.72(3) = 2.16
(9, 6)   1.20(6) + 0.72(9) = 13.68
(12, 4)   1.20(4) + 0.72(12) = 13.44
(16, 0)   1.20(0) + 0.72(16) = 11.52

Maximized at 9 cases of generic
peaches and 6 cases of fancy
peaches

REVIEW PROBLEMS, PAGES 123–124

1. (a)

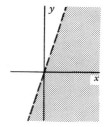

(b)

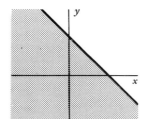

(c)

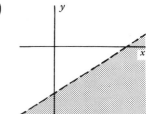

1. (d)

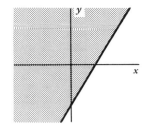

2. (a)

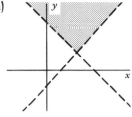

2. (b)

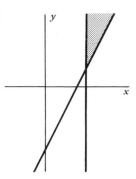

2. (c)

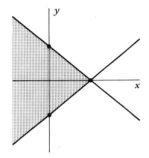

(d) no solutions

3.

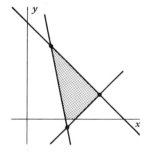

4.

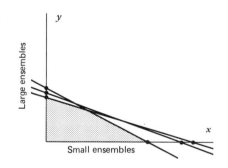

CHAPTER 8

EXERCISES 8-A, PAGES 129–130

I. **1.** $2x^2 + 5xy - 3y^2$
   5. $r^8 + 2r^6 - r^5 - r^3 - 2r + 1$
   9. $2x^6 - 8x^5 + 4x^4 - 18x^3 + 8x^2$

II. **1.** $6x^2 + 19x + 10$
    5. $3x^2 + 5xy - 2y^2$
    9. $x^2 + 8x + 16$
   13. $x^4 - 2x^2y^2 + y^4$

3. $3x^4 + x^3 + 7x^2 + 2x + 2$
7. $x^3 + x^2y - xy^2 + 2y^3$

3. $71x^2 - 139x - 6$
7. $4x^2 - y^2$
11. $16y^2 - 16y + 4$

III. 1. $4xy$ **3.** $2x^2 - 2y^2$

IV. 1. $c = 2, d = 6$ **3.** $c = 2, d = -1$

   5. $c = 5, d = -5$ or $c = -5, d = 5$

V. 1. (a) $(x + 3)(x + 3)$ **(b)** $(2x + 3)(x + 3)$

   3. $\pi(25 - x^2)$ or $25\pi - x^2\pi$

EXERCISES 8-B, PAGE 133

1. $2x + 1, 0, 2x^2 - 7x - 4 = (2x + 1)(x - 4) + 0$

3. $2x^2 - x + 1, 0, 2x^4 - x^3 + 7x^2 - 3x + 3 = (2x^2 - x + 1)(x^2 + 3) + 0$

5. $x^2 + x + 1, 0, x^3 - 1 = (x^2 + x + 1)(x - 1) + 0$

7. $0, x^3 - 1, x^3 - 1 = 0 \cdot x^5 + (x^3 - 1)$

9. $x^2 + 5x + 6, 0, x^4 - 13x^2 + 36 = (x^2 + 5x + 6)(x^2 - 5x + 6) + 0$

EXERCISES 8-C, PAGES 139–141

I. 1. $1 \cdot 12, 2 \cdot 6, 3 \cdot 4, (-1)(-12), (-2)(-6), (-3)(-4)$

   3. $1 \cdot 63, 3 \cdot 21, 9 \cdot 7, (-1)(-63), (-3)(-21), (-9)(-7)$

   5. $1 \cdot 36, 2 \cdot 18, 3 \cdot 12, 4 \cdot 9, 6 \cdot 6, (-1)(-36), (-2)(-18), (-3)(-12), (-4)(-9), (-6)(-6)$

II. 1. $2 \cdot 7 \cdot 59$ **3.** $2^4 \cdot 263$ **5.** $11 \cdot 13 \cdot 17$

III. 1. $4(x + 2y)$ **3.** $ab^2(-ab + 2)$ **5.** $(x + 2y)(x - 3y)$

   7. $(2x - 1)(3x + 1)$ **9.** $b(x - y)(x^2 + 7xy + y^2)$

   11. $b(x - y)^2(1 - 5b)(1 + 5b)$ **13.** $(y + x)(x - 1)$

   15. $(x - y)(x + y)(x - 2y)$

IV. 1. $(x + 5)(x + 3)$ **3.** $(3x + 2)(x + 1)$

   5. $(x + 5)(x + 1)$ **7.** $-(x + 9)(x - 6)$

   9. $(2x + 3)(x + 5)$ **11.** $-(2x + 3)(3x - 1)$

   13. Irreducible   **15.** Irreducible

   17. $-(4x + 3)(3x - 1)$ **19.** $(10x - 3)(5x - 1)$

   21. Irreducible

V. 1. $(x - 7)(x + 7)$ **3.** $(5 - 2n)(5 + 2n)$

   5. $(r - 6)(r + 6)$ **7.** $(x^2 + 9)(x - 3)(x + 3)$

   9. No

VI. 1. $(x - 2)(x^2 + 2x + 4)$ **3.** $(5y + 2)(25y^2 - 10y + 4)$

   5. $5(2a + 5b)(4a^2 - 10ab + 25b^2)$

   7. $(1 - x - y)(1 + x + y + (x + y)^2)$

VII. 1. $(x + 2)(x + 2)(x - 3)(x + 1)$ **3.** $c = 12$

   5. 40 feet per second

VIII. 1. $2(3x + 2)(2x - 3)$ **3.** $2x(3x - 2)(x + 1)$

   5. $(y - 2)(y + 2)(y - 2)(y + 2)$ **7.** $n(n - 2)(n - 1)$

   9. $x^2(4 - 5x)(4 + 5x)$ **11.** $(x - y)(x + y)(x^2 + xy + y^2)(x^2 - xy + y^2)$

   13. $2(2x + 5)(2x - 5)$ **15.** $x(x - 6)(x + 6)$

   17. $(2x + yz)(4x^2 - 2xyz + y^2z^2)$ **19.** $(x + y - 1)[(x + y)^2 + x + y + 1]$

REVIEW PROBLEMS, PAGES 141–142

1. (a) $4x^2 + 11xy - 3y^2$

   (c) $5x^4 - 2x^3 + 6x^2 - 2x + 1$

(b) $9y^2 - 12y + 4$

2. (a) $5x^2 - 3y^2$

(b) $-12xy$

3. $x + 3$

4. (a) $(x - 8)^2$

   (c) $(5x - y^2)(5x + y^2)$

   (e) $(2a + 5 - b)(2a + 5 + b)$

   (g) Polynomial is irreducible

(b) $b(b + 4)(b - 5)$

(d) $-2x^2y(5xy + y - 10)$

(f) $(2y + 3)(4y^2 - 6y + 9)$

(h) $(4x^2 - y)(16x^4 + 4x^2y + y^2)$

5. $(x + 6)(y + 2)$

6. $1 \cdot (x - 2)^2$

CHAPTER 9

EXERCISES 9-A, PAGE 145

1. $\dfrac{150}{x + 20}$

3. $\dfrac{15}{x + 2} + \dfrac{15}{x - 2}$

5. $\dfrac{143}{w + 3}$

EXERCISES 9-B, PAGES 147–148

I. 1. $\dfrac{1}{2}, \dfrac{4}{3}, \dfrac{4}{15}, 0$

3. Undefined, 0, 6, -4

5. 1, undefined, 4, $\dfrac{1}{7}$

II. 1. 0

3. 0, -3

5. Defined for all values of x

7. $x = -\dfrac{1}{5}$

9. $x = -3, x = -8$

III. 1. -2 **3.** 3 **5.** 0, 1 **7.** None **9.** None

IV. For a fraction to be undefined, the denominator must be zero. For a fraction to be zero, the numerator must be zero but the denominator a nonzero value.

EXERCISES 9-C, PAGES 150–151

I. 1. $\dfrac{x^2 + x}{x^3}, \dfrac{x^2 - 1}{x^3 - x}, \dfrac{2x + 2}{2x^2}$

3. $\dfrac{x^4 - 8x}{x^3 + 4x}, \dfrac{2x^3 - 16}{2x^2 + 8}, \dfrac{x^5 - 8x^2}{x^4 + 4x^2}$

5. $\dfrac{2x}{5x^2}, \dfrac{10}{25x}, \dfrac{4}{10x}$

II. 1. Yes **3.** Yes **5.** No **7.** No **9.** No

III. 1. x^4 **3.** y^{15} **5.** 2 **7.** $3x^2$ **9.** $\dfrac{1}{y^5}$

IV. 1. $\dfrac{x}{x^3 - 1}$ **3.** $\dfrac{x - 2}{x + 2}$ **5.** $\dfrac{x^3 - 9x}{3}$

7. $\dfrac{x - 6}{x - 5}$ **9.** $\dfrac{(x - 1)(x + 2)}{(x + 1)(x - 2)}$

EXERCISES 9-D, PAGES 154–155

I. 1. $\dfrac{x^2 + 2x}{x^2 + 4x + 3}$ for $x \neq -1, -3$ **3.** $\dfrac{21x^3 + 14x^2}{5x + 4}$ for $x \neq -\dfrac{4}{5}$

5. $\dfrac{x^3}{3x^5 - x^3 - 3x^2 + 1}$ for $x \neq 1$ **7.** $\dfrac{x^3 + 2x^2}{x^3 + 8x^2 + 19x + 12}$ for $x \neq -1, -3, -4$

9. $\dfrac{3}{5x^2}$ for $x \neq 0$ **11.** $\dfrac{x - 1}{3x + 3}$ for $x \neq 0, -1, 1$

13. 1 for $x \neq 1, -\dfrac{3}{2}$ **15.** $\dfrac{(x^2 - 3x - 4)^2}{x^{13}}$ for $x \neq 0, -1, 4$

17. $\dfrac{5(x + 3)^2}{x(3x + 1)}$ for $x \neq 0, -\dfrac{1}{3}, -3$ **19.** $\dfrac{(x - 2)^2}{x^6}$ for $x \neq 0, 2, 5$

II. 1. 0 **3.** 0

5. 0, undefined, undefined, 0

EXERCISES 9-E, PAGES 159–160

I. 1. $\dfrac{3x + 4}{x^3}$ for $x \neq 0$ **3.** $\dfrac{2x + 1}{x^2}$ for $x \neq 0$

5. $\dfrac{x^4 + 4x^2 + 4x + 1}{2x^3 + x^2}$ for $x \neq 0, -\dfrac{1}{2}$

7. $\dfrac{3x^2 - 1}{x^3 - x}$ for $x \neq 0, 1, -1$ **9.** $\dfrac{x^3 - x^2 + 2x - 1}{x^2 - x}$ for $x \neq 1, 0$

11. $\dfrac{x^2 - 3x}{2x + 3}$ for $x \neq -\dfrac{3}{2}$ **13.** $\dfrac{3x^2 + 3}{2x^2 - 5x + 2}$ for $x \neq \dfrac{1}{2}, 2$

15. $\dfrac{3x^2}{4x^2 - 1}$ for $x \neq \dfrac{1}{2}, -\dfrac{1}{2}$

II. 1. $\dfrac{x(5x^2 + 1)}{(x^2 - 1)^2}$ **3.** $\dfrac{x^3 - 7x^2 - 22x - 4}{(x + 1)(x + 3)(x - 4)}$

5. $\dfrac{x^2 + 1}{x}$ **7.** $\dfrac{x^2 + 1}{x^2 + x}$

III. 1. 0 **3.** $\dfrac{1}{3}$ **5.** $\dfrac{5}{2}$

IV. 1. 0 **3.** $\dfrac{1}{6}$ **5.** $\dfrac{5}{2}$

V. 1. 6 **3.** Undefined

REVIEW PROBLEMS, PAGE 161

1. (a) $\dfrac{3}{x^4}$ **(b)** $\dfrac{x-3}{x+4}$

2. (a) $\dfrac{1}{5}$, 0, 1, 2, $\dfrac{25}{17}$ **(b)** $\dfrac{11}{30}$, not defined, $\dfrac{1}{10}$, not defined, $\dfrac{9}{10}$

     (c) 4, 1, not defined, 3

3. (a) $\dfrac{(x+3)(x-1)}{(x+1)(x-3)}$ **(b)** $\dfrac{(x-2)^2}{(x+1)^2}$

     (c) $\dfrac{x^8}{(x^2-1)(x^2+1)}$ **(d)** $\dfrac{2(x+1)}{3(x-1)}$

4. (a) $\dfrac{x^2+1}{2x}$ **(b)** $\dfrac{4x^2+2}{(x-1)^2(x+1)}$ **(c)** $\dfrac{-2}{(x+2)(x+4)}$

5. $\dfrac{19}{18}$

6. (a) $\dfrac{-2}{x(x+2)}$ **(b)** $\dfrac{-1}{x(x-1)}$

CHAPTER 10

EXERCISES 10-A, PAGES 169–170

1. (a) 13   **(b)** 1.73   **(c)** 1.73   **(d)** Impossible
     (e) 3.46   **(f)** 2.24

3. (a) $\dfrac{1}{2}$ **(b)** 1   **(c)** $1+\sqrt{7}$ **(d)** Impossible

5. (a) 37   **(b)** $10\sqrt{2}$ **(c)** $2\sqrt{5}$ **(d)** $\sqrt{6/4}$

7. 10, 24 and 26
9. $\sqrt{71} \times \sqrt{71}$
11. (a) 8   **(b)** 8

EXERCISES 10-B, PAGES 175–176

I. 1. 2   **3.** 2   **5.** 2   **7.** 2   **9.** 2   **11.** 0

II. 1. $-1, -4$ **3.** $\dfrac{4}{3}, -1$ **5.** $\dfrac{7}{2}, -\dfrac{7}{2}$ **7.** $-\dfrac{5}{2}, \dfrac{1}{3}$

     9. $6, -\dfrac{3}{4}$ **11.** No real solutions

III. 1. 5×40 **3.** $9, -4$ **5.** 4   **7.** 4

IV. 1.

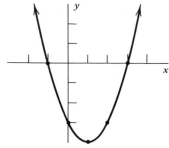

Graph touches x axis at
$(-1, 0)$ and $(3, 0)$
$x = -1$ and $x = 3$

3.

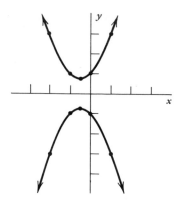

5. $a = -\dfrac{8}{3}, c = \dfrac{20}{3}$, yes

EXERCISES 10-C, PAGE 178

I. 1. (a) 2 pts

(c) $\left(-\dfrac{1}{2}, -\dfrac{9}{4}\right)$

(b) $(1, 0), (-2, 0)$

(d) $(0, -2)$

3. (a) 2 pts

(c) $(0, 9)$

(b) $(3, 0), (-3, 0)$

(d) $(0, 9)$

5. (a) 0 pts

(c) $(0, 1)$

(b) ——

(d) $(0, 1)$

II. $-\dfrac{b}{2a}$

EXERCISES 10-D, PAGES 181–182

I. 1. $\dfrac{2}{3}$

3. 2

5. $-\dfrac{11}{9}$

7. $\dfrac{7 + \sqrt{73}}{2}, \dfrac{7 - \sqrt{73}}{2}$

9. $\dfrac{2}{3}$

11. -4

13. No solution

II. 1. $\dfrac{1}{2}$ mile per hour

3. 2 hours

5. $r_{\text{boat}} = 10, r_{\text{plane}} = 510$

REVIEW PROBLEMS, PAGES 182–183

1. (a) $16a - b$

(b) $a\sqrt{3a}$

(c) $x^4 y^5$

2. $2\sqrt{13}$

3. $\sqrt{\dfrac{25}{2}} = \dfrac{5}{\sqrt{2}}$

4. (a) $x = 2/5, x = -3$

(c) $x = 6$

(b) No real number solutions

(d) $x = \sqrt{2}, x = -\sqrt{2}$

5. (a) $x = -1$, $x = 3$, $(1,4)$

(b) $x = \dfrac{-6 + \sqrt{6}}{2}$, $x = \dfrac{-6 - \sqrt{6}}{2}$, $(-3, -3)$

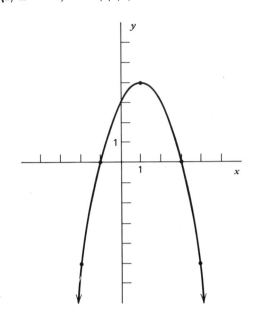

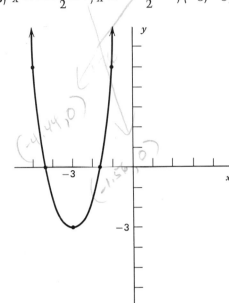

6. 130

7. walking speed: 3 mph
bus speed: 30 mph

8. 56 in. by 90 in.

CHAPTER 11

EXERCISES 11-A, PAGES 189–190

I. 1. $\dfrac{7}{55}$ **3.** $\dfrac{8}{5}$ **5.** 1600   **7.** $\dfrac{25}{4}$ **9.** $\dfrac{1}{50}$

II. 1. A varies directly with r^2; $k = \pi$
   3. D varies jointly with r and t; $k = 1$
   5. y varies inversely with x^2; $k = 1$
   7. r varies directly with t^2; $k = 100$
   9. x varies inversely with y; $k = 9/2$

III. 1. 360 ft.   **3.** approximately 22 minutes
   5. 10,000 decibels

EXERCISES 11-B, PAGES 194–195

I. 1. 1   **3.** $\dfrac{1}{512}$ **5.** 1   **7.** $\dfrac{1}{25}$ **9.** 1

II. 1. x^2 **3.** $\dfrac{1}{x^4}$ **5.** x^3 **7.** u^2 **9.** $\dfrac{b^3}{a^3}$

11. $\dfrac{1}{a^{12}}$ **13.** $a^4 b^2$ **15.** $\dfrac{a^8}{b^8}$ **17.** x^{16} **19.** $\dfrac{c^{12}}{a^4 b^8}$

21. xy^5 **23.** $\dfrac{yz^3}{x^3}$ **25.** $\dfrac{x^2}{y^2}$ **27.** $\dfrac{b}{a^{19}}$ **29.** $\dfrac{y}{2 + 5y}$

31. $x(x + 1)^2$

REVIEW PROBLEMS, PAGES 195–196

1. (a) $k = \dfrac{1}{7}$ **(b)** $k = 45$ **(c)** $k = 64$

2. (a) x varies inversely with y^2; the constant of variation is 5.

    (b) t varies directly with w and inversely with z; the constant of variation is 2.

3. 18 people

4. (a) $\dfrac{1}{49}$ **(b)** $-\dfrac{1}{8}$ **(c)** $\dfrac{4}{25}$

5. (a) $\dfrac{1}{x}$ **(b)** a **(c)** $\dfrac{x^6}{y^3}$ **(d)** $\dfrac{z^6}{x^8 y^2}$

CHAPTER 12

EXERCISES 12-A, PAGES 201–202

I. 1.

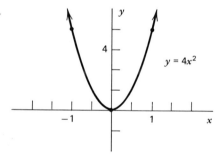

$y = 4x^2$

3.

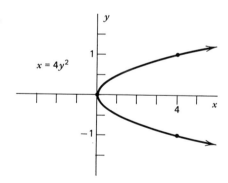

$x = 4y^2$

II. 1. $y = .00075x^2$

EXERCISES 12-B, PAGE 203

I. 1. $x^2 + y^2 = 49$ **3.** $(x - 2)^2 + (y + 9)^2 = 9$

II. 1. $(-1, 2)$, $r = \sqrt{5}$

3. $(0, 0)$, $r = 3$

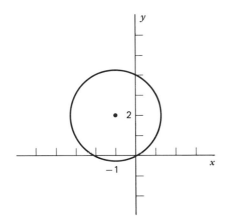

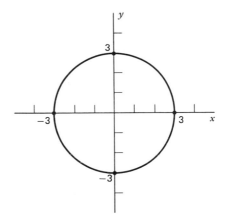

5. $(0, 0)$, $r = 2$

7. $(1, -3)$, $r = 4$

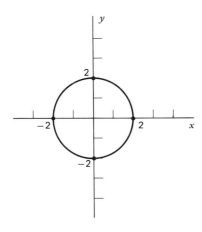

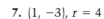

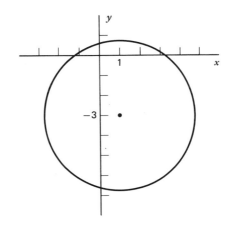

EXERCISES 12-C, PAGE 208

I. 1. $(0, 0)$, $(2\sqrt{7}, 0)$, $(-2\sqrt{7}, 0)$

   3. $(-3, -1)$, $(-3 + 2\sqrt{6}, -1)$, $(-3 - 2\sqrt{6}, -1)$

II. 1. $8, 6, 2\sqrt{7}$

3. $7, 5, 2\sqrt{6}$

III. 1. $\dfrac{x^2}{64} + \dfrac{y^2}{36} = 1$

3. $\dfrac{(x + 3)^2}{49} + \dfrac{(y + 1)^2}{25} = 1$

IV. 1. $\dfrac{x^2}{25} + \dfrac{y^2}{16} = 1$

3. $(x + 3)^2 + \dfrac{(y - 1)^2}{4} = 1$

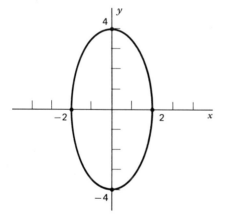

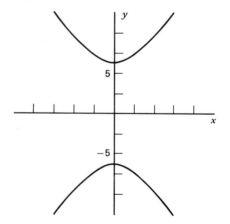

EXERCISES 12-D, PAGES 213–214

I. 1. ellipse

3. hyperbola

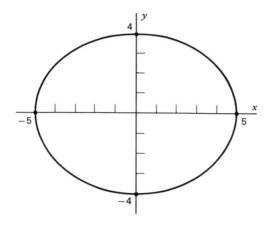

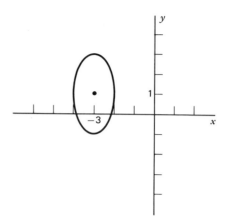

5. circle

7. hyperbola

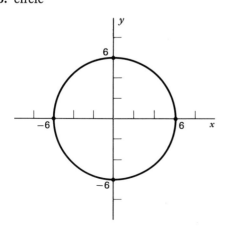

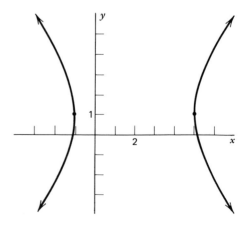

9. ellipse

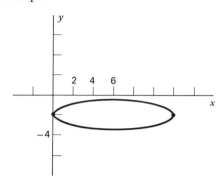

11. circle

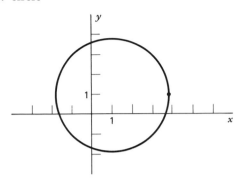

13. ellipse

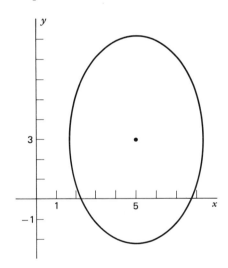

II. 1. (a) $y = -\frac{4}{5}x$ $y = \frac{4}{5}x$ **(b)** $y = -\frac{4}{3}x$ $y = \frac{4}{3}x$

(c) $y = -\frac{1}{2}x$ $y = \frac{1}{2}x + 6$

3. $\dfrac{(x-1)^2}{9} - \dfrac{(y-2)^2}{36} = 1$

EXERCISES 12-E, PAGES 217–218

I. 1. 3   **3.** 1   **5.** -16 **7.** $-\dfrac{11}{4}$

9. $2t - 20t^2 - 2$

II. 1. $12, 2, -\dfrac{1}{4}, 2, 12$

3. no

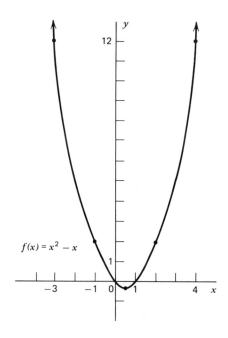

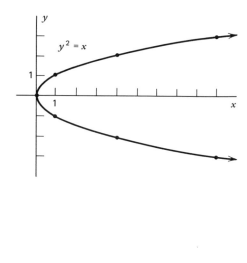

5. No, the vertical line $x = a$ is not a function.

7. $y = \sqrt{16 + 4x^2}$ and $y = -\sqrt{16 + 4x^2}$

REVIEW PROBLEMS, PAGES 219–220

1. (a) $(x - 2)^2 + (x - 3)^2 = 4$

(b) $\dfrac{(x + 2.5)^2}{3.5^2} + \dfrac{(y + 2)^2}{4} = 1$

(c) $\dfrac{x^2}{1} - \dfrac{y^2}{4} = 1$

2.

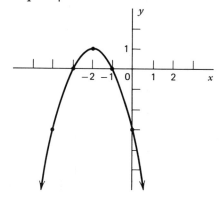

3. (a)

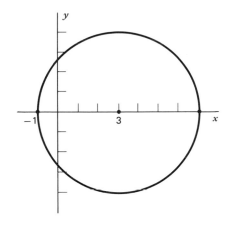

(b)

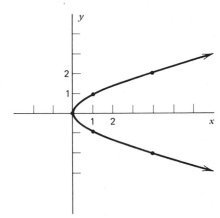

(c)

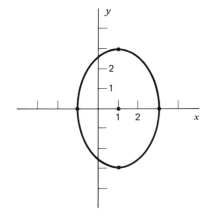

(d)

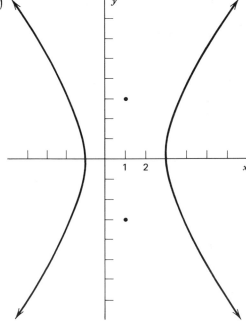

4.

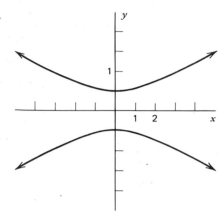

$$y = \sqrt{1 + \frac{x^2}{4}}$$

$$y = -\sqrt{1 + \frac{x^2}{4}}$$

5.

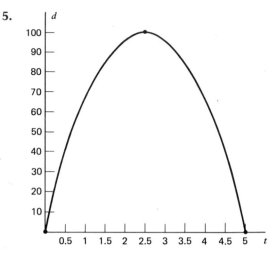

APPENDIX A

EXERCISES A-1, PAGES 228–230

I. 1. 18   **3.** 8   **5.** 9   **7.** 9

II. 1. 32   **3.** 56   **5.** 32

III. 1. All have bases and altitudes of the same length.

    2. $(a + b)^2$ is the area of the whole square. The four parts of the square have areas a^2, ab, b^2, and ba. Thus $(a + b)^2 = a^2 + ab + ba + b^2 = a^2 + 2ab + b^2$.

EXERCISES A-2, PAGES 232–233

I. 1. 20   **3.** 20   **5.** no

II. 1. 18   **3.** 22   **5.** 24

III. 1. (a)

(b)

    (c) no

3. 32

APPENDIX B

EXERCISES B-1, PAGES 240–241

I. **1.** 32   **3.** 30   **5.** 10   **7.** 2
   9. 6   **11.** 25

II. **1.** $4 + (6 \div 3)$ **3.** $(8 \div 2) \times 2$

   5. $\dfrac{8}{3 + 1}$ **7.** $(4 + 7) \times 6$

   9. $-5 + 9$

EXERCISES B-2, PAGE 243

I. **1.** 43   **3.** 803   **5.** 400   **7.** 147

II. **1.** $7^2 - 2$ **3.** $4^{(6-4)}$ **5.** $9 + 3^2$ **7.** $(3^2)^2$

APPENDIX C

EXERCISES C-1, PAGES 247–248

I. **1.** 24   **3.** 4.35   **5.** 65.19   **7.** 0.013
   9. 0.7

II. **1.** John's salary
   3. The amount of the loan
   5. The number of voters registered last year
   7. The number of books borrowed from the library
   9. The total wages of people working in the city

EXERCISES C-2, PAGES 249

1. 6.5   **3.** 7.74   **5.** 75%   **7.** 96   **9.** 344%
11. 200   **13.** 64.9351   **15.** .48   **17.** 200   **19.** 0.45

EXERCISES C-3, PAGES 251–252

1. $1.13 per hour   **3.** $14,700
5. 3.6%   **7.** $14,500
9. 29.8%   **11.** 32.25%

APPENDIX D

EXERCISES D-1, PAGES 261–263

II. 1. $\dfrac{2}{6}, \dfrac{3}{9}, \dfrac{4}{12}, \dfrac{5}{15}, \dfrac{6}{18}$ **3.** $\dfrac{10}{16}, \dfrac{15}{24}, \dfrac{20}{32}, \dfrac{25}{40}, \dfrac{30}{48}$

5. $\dfrac{54}{63}, \dfrac{18}{21}, \dfrac{6}{7}, \dfrac{12}{14}, \dfrac{24}{28}$ **7.** $\dfrac{6a}{4b}, \dfrac{9a}{6b}, \dfrac{3a^2}{2ab}, \dfrac{3ab}{2b^2}, \dfrac{12a}{8b}$

III. 1. $\dfrac{5}{40}$ **3.** $\dfrac{7}{6}$ **5.** $\dfrac{8}{13}$

IV. 1. 0.6 **3.** 1.8 **5.** .6842 **7.** 6 **9.** 5.5333

11. 3.3

V. 1. = **3.** < **5.** = **7.** < **9.** >

VI. 1. $\dfrac{7}{13}$ **3.** $\dfrac{4}{5}$ **5.** $\dfrac{71}{132}$ **7.** $\dfrac{16}{25}$

VII. 1. $\dfrac{24}{40}$ **3.** $\dfrac{5}{8}$

EXERCISES D-2, PAGES 266–267

I. 1. $\dfrac{4}{2}, \dfrac{6}{3}, \dfrac{8}{4}, \dfrac{10}{5}, \dfrac{12}{6}$ **3.** $\dfrac{30}{20}, \dfrac{6}{4}, \dfrac{9}{6}, \dfrac{12}{8}, \dfrac{15}{10}$

5. $\dfrac{40}{70}, \dfrac{8}{14}, \dfrac{12}{21}, \dfrac{16}{28}, \dfrac{20}{35}$ **7.** $\dfrac{10}{2}, \dfrac{20}{4}, \dfrac{15}{3}, \dfrac{25}{5}, \dfrac{30}{6}$

9. $\dfrac{70}{90}, \dfrac{14}{18}, \dfrac{21}{27}, \dfrac{28}{36}, \dfrac{35}{45}$ **11.** $\dfrac{100}{128}, \dfrac{250}{320}, \dfrac{1000}{1280}, \dfrac{50}{64}, \dfrac{500}{640}$

II. 1. $\dfrac{7}{3}$ **3.** $\dfrac{27}{5}$ **5.** $\dfrac{33}{8}$ **7.** $\dfrac{7}{3}$ **9.** $\dfrac{23}{8}$

III. 1. $1\dfrac{2}{5}$ **3.** 2 **5.** $4\dfrac{21}{25}$ **7.** $4\dfrac{1}{4}$ **9.** $2\dfrac{2}{3}$

EXERCISES D-3, PAGE 270

I. 1. > **3.** < **5.** < **7.** < **9.** >

II. 1. $\dfrac{2}{23}, \dfrac{1}{11}, \dfrac{3}{31}, \dfrac{1}{9}, \dfrac{2}{10}$ **3.** $\dfrac{4}{5}, \dfrac{7}{8}, \dfrac{15}{16}, \dfrac{17}{18}, \dfrac{21}{22}$

EXERCISES D-4, PAGES 279–281

I. 1. $\dfrac{5}{12}$ **3.** 1 **5.** $\dfrac{3}{10}$ **7.** $\dfrac{3}{20}$ **9.** $\dfrac{7}{18}$

II. 1. $\dfrac{9}{10}$ **3.** $\dfrac{1}{2}$ **5.** 6 **7.** $\dfrac{121}{3}$ **9.** $\dfrac{121}{36}$

III. **1.** $\frac{2}{3}$ **3.** 1   **5.** 7   **7.** $\frac{1}{6}$ **9.** $\frac{9}{2}$

11. $\frac{1}{12}$ **13.** $\frac{65}{18}$ **15.** $\frac{16}{9}$ **17.** $\frac{15}{4}$ **19.** $\frac{45}{64}$

IV. **1.** $\frac{7}{6}$ **3.** $\frac{17}{15}$ **5.** $\frac{333}{385}$ **7.** $\frac{25}{6}$ **9.** $\frac{247}{24}$

11. $\frac{1}{6}$ **13.** $\frac{1}{72}$ **15.** $\frac{3}{32}$ **17.** $\frac{5}{4}$ **19.** $\frac{47}{12}$

V. **1.** $\frac{13}{24}$ **3.** $\frac{1}{5}$ **5.** $\frac{121}{32}$

VI. **1.** $\frac{13}{24}$ **3.** $\frac{1}{5}$ **5.** $\frac{121}{32}$

VII. The distributive property.

EXERCISES D-5, PAGES 283–284

1. $\frac{1}{8}$ **3.** $\frac{13}{12}$ **5.** 4   **7.** $\frac{5}{6}$ **9.** $-\frac{17}{24}$

GLOSSARY

Algebraic Expression
An algebraic expression is formed when numbers and variables are combined using addition, subtraction, multiplication, division, or the operation of raising to a power. *Examples:*
$$x^2 + 4y, \quad x - \frac{1}{2}\sqrt{x}, \quad \frac{x + 2y}{x^2}.$$

Associative Property
Addition is associative because $a + (b + c) = (a + b) + c$ for any numbers a, b, and c. Multiplication is associative because $a \cdot (b \cdot c) = (a \cdot b) \cdot c$ for any numbers a, b, and c. Subtraction and division are not associative.

Asymptote
Some graphs are shaped almost like straight lines at their extremes. A line that a graph approaches is called an asymptote.

Axis (Axes)
See **Coordinate System.**

Base
See **Exponent.**

Circumference
The distance around a circle is called its circumference. The circumference is equal to the product of the number π and the diameter of the circle.

Coefficient
When a number is multiplied by a power of a variable, the number is called the coefficient of the power of the variable. For example, 3 is the coefficient of x^4 in the expression $3x^4$.

Commutative Property
Addition is commutative because $a + b = b + a$ for any numbers a and b. Multiplication is commutative because $a \cdot b = b \cdot a$ for any numbers a and b. Subtraction and division are not commutative.

Conic Section(s)
The curves that can be formed by cutting a double cone with a plane are called conic sections. Parabolas, circles, ellipses, and hyperbolas are conic sections.

Constant
When a number occurs as a term in an expression with variables, the number may be called a constant because it does not change. For example, 200 is a constant in the expression $x^7 + 200$.

Constant of Variation
See **Variation.**

Coordinate System
Two intersecting number lines (each called an *axis; axes* is plural) can be used to assign a pair of numbers to each point in a plane. The most common coordinate system has a horizontal axis (x axis) and a vertical axis (y axis). The two numbers assigned to a point are called its *coordinates.* The point with coordinates $(0, 0)$ is called the *origin* of the coordinate system.

Cross-Multiplication
See **Equivalent Fractions.**

Degree of Polynomial
See **Polynomial.**

Distributive Property
Multiplication distributes over addition because $a \cdot (b + c) = (a \cdot b) + (a \cdot c)$ and $(b + c) \cdot a = (b \cdot a) + (c \cdot a)$ for any numbers a, b, and c.

Equation
A statement that two algebraic expressions are equal is an equation. *Examples:* $x^2 = 2x + 8$; $\frac{1}{x^3} = y$. Numbers that give a true statement when they replace the variables in an equation are called *solutions* to the equation. Both 4 and -2

are solutions to $x^2 = 2x + 8$ because $4^2 = 2(4) + 8$ and $(-2)^2 = 2(-2) + 8$.

Equivalent Equations

Two equations with exactly the same solutions are said to be equivalent; $2x = 6$ and $2x - 5 = 1$ are equivalent because each has as a solution the number 3 and only the number 3.

Equivalent Fractions

Two fractions that represent the same number are said to be equivalent (or equal). One way to obtain equivalent fractions is to multiply the numerator and denominator of a fraction by the same number; for example, $\frac{2}{3} = \frac{10}{15}$. A check for equivalent fractions is to multiply the numerator of the first fraction by the denominator of the second and the denominator of the first by the numerator of the second. If these products are equal, the fractions are equivalent: $\frac{4}{6} = \frac{10}{15}$ because $4 \cdot 15 = 6 \cdot 10$. This procedure for checking equivalence is called *cross-multiplication*.

Exponent

In an expression like $(2x)^3$ the number 3 is called the *exponent* and $2x$ is called the *base*.

Factor

When an expression is written as a product, the expressions in the product are called factors. For example, because $2x^2 + x - 3 = (x - 1)(2x + 3)$, both $x - 1$ and $2x + 3$ are said to be factors of $2x^2 + x - 3$.

Factored Completely

See **Irreducible Polynomial**.

Function

When an equation in the variables x and y has the property that there is at most one y value corresponding to each x value, we say the equation describes y as a *function* of x.

Graph of an Equation

The graph of an equation is the collection of points whose coordinates are solutions to the equation.

Hypotenuse

In a right triangle, the side opposite the right angle is called the hypotenuse.

Inequality

A statement that one algebraic expression is less than another algebraic expression (or that one is greater than another) is called an inequality. *Example:* $x^2 < x + 12$. An inequality will sometimes be written with the symbol $\leq$ to indicate "less than or equal to" or $\geq$ to indicate "greater than or equal to." Numbers that give true statements when they replace the variables in an inequality are called *solutions* to the inequality. Any number between -3 and 4 is a solution to $x^2 < x + 12$.

Integers

The set of natural numbers and their opposites together with 0 is called the set of integers: $\{. . . , -3, -2, -1, 0, 1, 2, 3, ...\}$.

Irreducible Polynomial

A polynomial that cannot be written as a product of two factors of smaller, positive degree is said to be irreducible. A polynomial that has been written as a product of irreducible polynomials is said to be *factored completely*.

Least Common Denominator

The least common denominator of two fractions is the smallest number that is a multiple of both of the denominators. For example, the least common denominator for $\frac{5}{6}$ and $\frac{11}{15}$ is 30.

Linear Equation

A linear equation is an equation in which each term involving a variable has the form "a number times the variable." The equation $3x - 2 = x + 4$ is linear in one variable; the equation $y = 5x - 3$ is linear in two variables.

Natural Number

A natural number is one of the numbers 1, 2, 3, 4, ...

Order of Operations

In evaluating an expression, it is assumed, in the absence of parentheses, that exponentiation is performed first, then multiplication and division in order left to right, and finally addition and subtraction in order from left to right.

Origin

See **Coordinate system**.

Parabola

The graph of a quadratic equation of the form $y = ax^2 + bx + c$ is a parabola that opens up if $a > 0$, and down if $a < 0$. The vertex of a parabola in this form has the x coordinate $\dfrac{-b}{2a}$.

Parallel Lines

Two lines that are called parallel if they do not intersect (even when extended indefinitely).

Perimeter of a Polygon

The perimeter of a polygon is the distance around the polygon. The perimeter is computed by adding the lengths of the line segments that make up the boundary of the polygon.

Perpendicular

Two lines are perpendicular if they form 90° angles at their point of intersection. A 90° angle is also called a *right angle*.

Polygon

A polygon is a two-dimensional geometric figure that has its boundary made up of line segments. Triangles, rectangles, and hexagons are examples of polygons.

Polynomial

A polynomial is an expression in the form $a_n x^n + \ldots + a_2 x^2 + a_1 x^1 + a_0$ where the coefficients $a_n, \ldots, a_1, a_0$ can be any numbers and exponents of x are whole numbers. Examples: $\frac{1}{2}x^3 - x + 6$; $3x^2 - 49$. The *degree of a polynomial* is the largest exponent that appears in the polynomial; $\frac{1}{2}x^3 - x + 6$ is a polynomial of degree 3; $3x^2 - 49$ is a polynomial of degree 2.

Prime Number

A whole number greater than 1 that cannot be factored into a product of two whole numbers both less than the number itself is said to be a prime number.

Pythagorean Theorem

The Pythagorean theorem describes the relationship between the lengths of the three sides of a right triangle. It says that if c is the length of the hypothenuse and a and b the lengths of the two shorter sides, then $c^2 = a^2 + b^2$.

Quadratic Equation

A quadratic equation is an equation in which the expression in one of the variables is a polynomial of degree 2. For example, $3x^2 - 4x = 12$ is a quadratic equation in one variable; $y = x^2 + x + \frac{1}{4}$ is a quadratic equation in two variables.

Quadratic Formula

The quadratic formula gives the solutions to a quadratic equation in one variable in terms of the coefficients of the quadratic polynomial. The solutions to $ax^2 + bx + c = 0$ are $x = \dfrac{-b + \sqrt{b^2 - 4ac}}{2a}$ and $x = \dfrac{-b - \sqrt{b^2 - 4ac}}{2a}$.

Rational Number

A rational number is the quotient of two integers.

Rational Expression

A rational expression is a quotient of polynomials. The variable in a rational expression represents any number except numbers that make the denominator 0. In $\dfrac{2x}{x^2 - 1}$, for example, x cannot assume the value 1 or -1.

Real Number

A positive real number measures length; if b is a positive real number, $-b$ is also a real number; 0 is a real number.

Reciprocal

If $\dfrac{n}{m}$ is a rational number and $n \neq 0$, then $\dfrac{m}{n}$ is its reciprocal. If $\dfrac{P}{Q}$ is a rational expression and $P \neq 0$, then $\dfrac{Q}{P}$ is its reciprocal.

Right Angle
See **Perpendicular.**

Root

We speak of square roots, cube roots, fourth roots, and higher order roots. The square root of a number n is the positive number whose square is n (the square root of 16 is 4); the cube root of n is the number whose third power is n (the cube root of 8 is 2); the fourth root of n is the positive number whose fourth power is n (the fourth root of 81 is 3). This is the usual notation: $\sqrt{16} = 4$; $\sqrt[3]{8} = 2$; $\sqrt[4]{81} = 3$.

Simultaneous Solutions
See **System of Equations** and **System of Inequalities**.

System of Equations, System of Inequalities
Two or more equations or inequalities taken together make up a system. A solution to one equation (inequality) that is a solution to all the equations (inequalities) in a system is said to be a *simultaneous solution.*

Slope-Intercept Form
A linear equation that has the form $y = mx + b$ is said to be in slope-intercept form. In this form, the number m is the slope of the line described by $y = mx + b$, and b is its y intercept.

Slope of a Line
When the equation of a line is written in the form $y = mx + b$, the number m is the slope of the line. The slope of a line can be computed if two points of the line, (a, b) and (c, d) are known. In this case, $m = \dfrac{b - d}{a - c} = \dfrac{d - b}{c - a}$.

Solution of an Equation
See **Equation.**

Solution of an Inequality
See **Inequality.**

Square Root.
See **Root.**

Variable
A variable is a symbol, usually a letter, that stands for a number from a set of numbers. A variable permits us to write a statement about numbers without specifying each number in the statement.

Variation
If one variable equals a number times a second variable, we say the first variable *varies directly* with the second ($y = kx$ for some number k). If one variable equals a number divided by a second variable, we say the first variable *varies inversely* with the second $\left(y = \dfrac{k}{x}\right.$ for some number $k\left.\right)$. In both cases, the number k is called the *constant of variation.*

Whole Numbers
The numbers we count with, including 0, are called whole numbers. We can denote the set of whole numbers by $\{0, 1, 2, 3, \ldots\}$.

x Axis
See **Coordinate System.**

x Intercept
If the graph of an equation intersects the x axis at the number c on the x axis, then c is called an x intercept of the graph.

y Axis
See **Coordinate System.**

y Intercept
If the graph of an equation intersects the y axis at the number b on the y axis, the b is called a y intercept of the graph.

INDEX